Physical Reality and Common Sense

Physical Reality and Common Sense

*A Commonsense Description of
the Physical Reality Defined by
The Mathematical Models of
Relativity and Quantum Mechanics*

Walter G. Edwards

Library of Congress Control Number:		2011915444
ISBN:	Hardcover	978-1-4653-5814-1
	Softcover	978-1-4653-5813-4
	Ebook	978-1-4653-5815-8

This book was printed in the United States of America.

To order additional copies of this book, contact:
Xlibris Corporation
1-888-795-4274
www.Xlibris.com
Orders@Xlibris.com
102951

Contents

To Bette who listened and whose support made this possible

Preface

Like most students of physics, I had great difficulty in bridging the gap from the classical physics of Newton and Maxwell to the modern physics of relativity and quantum mechanics. The problem is caused by the fact that while classical physics is consistent with the physical reality paradigm generated by our sense experiences, the conventional interpretation of the mathematical models of relativity and quantum mechanics is not. The interpretation of modern physics presented in this book achieves my goal of establishing a commonsense description of physical reality that resolves the difficulty of understanding the mathematical models of modern physics.

The difficulty of comprehending the conventional interpretation of quantum mechanics is demonstrated by the fact that the developers of quantum mechanics believed that there is no quantum reality behind the mathematics of quantum mechanics.[p-1]

> Bohr once declared when asked whether the quantum mechanical algorithm could be considered as somehow mirroring an underlying quantum reality: "There is no quantum world. There is only an abstract quantum mechanical description. It is wrong to think that the task of physics is to find out how Nature is. Physics concerns what we can say about Nature."

> And for Heisenberg . . . in the experiments about atomic events we have to do with things and facts, with phenomena that are just as real as any phenomena in daily life. But the atoms or the elementary particles are not as real; they form a world of potentialities or possibilities rather than one of things or facts.

Feynman has more bluntly stated the problem of quantum reality.[p-2]

> I think I can safely say that nobody understands quantum
> mechanics . . . Do not keep saying to yourself, if you can possibly
> avoid it, "but how can it be like that?" because you will get "down
> the drain," into a blind alley from which nobody has yet escaped.
> Nobody knows how it can be like that.

Additionally, Diedrik Aerts has pointed out that "it is not at all clear 'what reality is' taking into account relativity theory."[p-3]

> Many textbooks on relativity theory give the impression that the
> theory is conceptually very well defined, contrary to textbooks on
> quantum mechanics, where it is usually openly admitted that many
> aspects of quantum theory are not understood at all. We think that
> it is actually possible to claim that also relativity theory, although
> seemingly very clearly put forward by Einstein himself, is not
> understood in many important aspects. *More concretely, it is not at
> all clear "what reality is" taking into account relativity theory.*

Finally, Stephen Hawking confirms that the conventional physics community has given up on trying to interpret the mathematical models of modern physics in terms of our sensory-based paradigm for external reality.[p-4]

> I take the positivist viewpoint that a physical theory is just a
> mathematical model and that *it is meaningless to ask whether it
> corresponds to reality*. All that one can ask is that its predictions
> should be in agreement with observation.

John Bell was one of the few physicists after Einstein who was concerned with the reality behind the mathematical models of modern physics. He believed that the older Lorentz preferred frame interpretation of special relativity theory described in chapter 2 is much more consistent with our sense experience–based perception of external reality.[p-5]

> I have for long thought that if I had the opportunity to teach this
> subject, I would emphasize the continuity with earlier ideas. Usually
> it is the discontinuity, which is stressed, the radical break with more

primitive notions of space and time. *Often the result is to destroy completely the confidence of the student in perfectly sound and useful concepts already acquired.*

Additionally, he believed that the de Broglie-Bohm pilot-wave synthesis of the particle *and* wave configurations of quantum objects described in chapter 5 is consistent with our sense experience–based paradigm for external reality, while the particle *or* wave configuration of the conventional Copenhagen interpretation of the mathematics of quantum mechanics is not. Bell summarized this approach in the following quote.[p-6]

> The de Broglie-Bohm synthesis, of particle and wave . . . combines quite naturally both the waviness of electron diffraction and interference patterns, and the smallness of individual scintillations, or more generally the definite nature of large scale happenings. The de B-B picture is also, by the way, quite deterministic. The initial configuration of the combined wave particle system completely fixes the subsequent development. That we cannot predict just where a particular electron will scintillate on the screen is just because we cannot know everything. That we cannot arrange for impact at a chosen place is just because we cannot control everything.

John Bell was an Irish physicist born in 1928. He spent most of his career at CERN in Geneva involved in quantum field theory. However, his greatest contribution to theoretical physics was Bell's inequality theorem (described in chapter 6), which is arguably one of the most important theorems of the twentieth century. Unfortunately, John Bell died unexpectedly of a stroke in 1990 before the acceptance of the now-known cosmic preferred frame structure of the universe established by the measurements of the cosmic microwave background radiation from the big bang origin of the universe.

We now know our universe contains over one hundred billion galaxies, and each one contains roughly one hundred billion stars. Furthermore, the measurements of the cosmic microwave background from the big bang origin of the universe establish a cosmic preferred frame for each region of the universe. Using the Doppler effect, measurements have been made that establish the velocity of the earth with respect to the cosmic preferred frame. I have used these results in chapter 4 to establish the coupling between local experiments and the total mass/energy in the external universe.

This extension of Bell's interpretation of the mathematics of modern physics is important because it is used to develop a commonsense description of physical reality that changes the foundations of modern physics by (1) restoring the *cause*-and-*effect* relationship between events in the real world, (2) eliminating the conventional belief that time travel to the past is possible, thus eliminating the paradoxes that have never and can never be resolved (e.g., killing one's own grandmother before your mother was born), (3) verifying Mach's principle that the centrifugal force in a rotating frame results from the motion with respect to the mass/energy in the external universe, thus eliminating the conventional belief in the possibility of an "empty" universe, (4) demonstrating that special relativity time dilation and gravitational time dilation are the same effect when referenced to the cosmic preferred frame, (5) using the de Broglie-Bohm pilot wave interpretation of quantum mechanics to describe quantum reality, and (6) eliminating the conflict between the conventional interpretation of special relativity and the nonlocality of quantum mechanics described in chapter 6.

The format used in this book is to first present direct quotes from respected physicists in the field to establish the description of the conventional interpretation of modern physics. Then, John Bell's preferred frame interpretation of special relativity and quantum mechanics is introduced by his direct quotes. In order to present this preferred frame interpretation of modern physics to the general public, there are no mathematical equations in the six chapters of the book. However, the mathematical analysis of the real experiments and thought experiments used to demonstrate the advantage of this approach are contained in the Appendix. Acceptance of this cosmic preferred frame interpretation of the mathematics of modern physics will require significant modifications to existing textbooks on relativity and quantum mechanics, and will verify that John Bell was one of the most important physicists of the twentieth century.

Walt Edwards   May 18, 2011

1.0 Classical Physics and the Perception of Physical Reality

Before addressing the difficult problems associated with comprehending the physical reality defined by the mathematics of relativity and quantum mechanics, it is important to establish the physical reality defined by our sense experience–based perception of the physical world. Any description of physical reality must be based on the definitions for (1) time, (2) space, (3) event, (4) reference frame, and (5) cause and effect. The mathematics of classical physics is a quantitative description of our commonsense paradigm for external reality.

Our commonsense paradigm for physical reality is based on combining the inputs to our five senses from the moment of birth. These experiences are stored in memory in the sequence they occur, which gives our perception of time. Time is measured with repeatable physical processes. The conventional method of measuring time is with a clock (i.e., each "tick" of the clock is repeatable). Newton's *absolute time* that is the same anywhere in the universe is the only time defined in classical physics. The concept of space evolves from a combination of seeing distant objects and then moving to touch them. The conventional way of measuring space (distance) is with a measuring rod. The length of a measuring rod is assumed to be the same anywhere in the universe and measures distance in Newton's *absolute space*.

An event in classical physics is a specific occurrence related to objects in physical reality (e.g., light beam hitting a mirror, ball hitting the ground, etc.). The events described by the mathematics of classical physics must be referenced to a frame containing three space distances and a time standard. A typical xyzt reference frame is shown in figure 1-1. Here, the x, y, and z values represent the three distances from the origin, while the t value represents a universal time that is the same for all points in the reference frame.

Einstein has stressed the requirement for cause-and-effect relationships to comprehend the physical reality behind the mathematical models of physics.[1-1]

> I am standing in front of a gas range. Standing alongside of each other on the range are two pans so much alike that one may be mistaken for the other. Both are half full of water. I notice that steam is being emitted continuously from the one pan, but not from the other. I am surprised at this, even if I have never seen either a gas range or a pan before. But if I now notice a luminous something of bluish colour under the first pan but not under the other, I cease to be astonished, even if I have never before seen a gas flame. For I can only say that this bluish something will cause the emission of the steam, or at least *possibly* it may do so. *If, however, I notice the bluish something in neither case, and if I observe that the one continuously emits steam whilst the other does not, then I shall remain astonished and dissatisfied until I have discovered some circumstance to which I can attribute the different behaviour of the two pans.*

Additionally, our knowledge of the evolution of earth and its life forms over the last four and a half billion years of cosmic time (e.g., rock formations, fossil records, carbon dating, etc.) supports the fact that the physical processes described by the mathematical models of classical and modern physics all existed in the universe in their present form independent of our perception of them, long before humans were on the earth. John Bell supports this view that classical physical reality as well as quantum reality is observer independent.[1-2]

> There is indeed much talk of "observables" in quantum theory books. And from some popular presentations the general public could get the impression that the very existence of the cosmos depends on our being here to observe the observables. I do not know that this is wrong. I am inclined to hope that we are indeed that important. But I see no evidence that it is so in the success of contemporary quantum theory.

This commonsense description of the physical reality of classical physics given here establishes the concepts to be used to develop the physical reality defined by the mathematical models of relativity and quantum mechanics.

1.1   The Metric System Standards

In order to construct a mathematical model to relate the dynamics of physical events, a frame of reference, a distance interval standard, and a time interval standard must all be established. The international unit system used throughout the scientific community (SI units) is based on the metric system, where the *meter* was meant to be one ten-millionth part of a meridian from the equator to the pole, while the *second* was defined as a fraction (1/31,556,925,9747) of the tropical year 1900. These definitions have since been replaced with the much more accurate and more easily constructed standards based on quantum mechanics. In 1960 the General Conference on Weights and Measures agreed to define the meter as a fixed number of wavelengths (1,650,763.73) of the orange-red spectral line in krypton 86, and in 1987 it was decided to define the second as fixed between the two hyperfine levels of the ground state of the atom of cesium 133. Using these two standards, the *velocity of light* is equal to 3.0×10^8 m/s.

The following metric system standards have been added here to demonstrate the additional parameters required to describe the reality of the universe.

The *kilometer* is equal to 1,000 m.

The *liter* is the basic unit of volume in the metric system. One liter is defined as the volume equal to one cubic decimeter (one-tenth of a meter). Therefore, one *milliliter* (mL) is equal to 1 cm^3 (1/100 of a meter).

The *gram* is the basic unit of mass in the metric system. One gram is defined as the mass of distilled water at 4°C contained in a cube whose edge is 1 cm (one-hundredth of a meter).

The Celsius temperature scale is the standard *temperature scale* in the metric system. A degree is defined as one-hundredth of the temperature difference between the *ice point* (melting point of pure ice—0°C) and the *steam point* (temperature of condensing steam under the pressure of one standard atmosphere—100°C).

1.2　　The Three Different Regions of Our Universe

The purpose of this section is to establish the different ways of doing physics in the three different regions of our universe. The *macro-world* region is roughly bounded by the diameter of an average human hair (10^{-4} m) and the distance to the moon (4×10^8 m). Classical physics and relativity theory were developed in this region of the universe. Here it is assumed that measurements can be made on macro objects that will not change the value of the parameters being measured. Also, it is important to realize the conventional interpretation of classical and modern physics is based on the belief that there is no coupling between this region and the external universe.

The *micro-world* region is roughly bounded by the diameter of an average human hair (10^{-4} m) and the diameter of the nucleus of an atom (4×10^{-15} m). Quantum mechanics was developed in this region of the universe. The conventional interpretation of quantum mechanics is that it is inconsistent with special relativity and there is no quantum reality. Also, it is believed that there is no coupling between this region and the external universe. Because measurements of quantum objects are made with other quantum objects, the parameters of the quantum objects being measured are changed.

The *external universe* region is roughly bounded by the distance to the moon (4×10^8 m) and the radius of the classical big bang model of the total universe (1.3×10^{26} m). This region is described by general relativity. It is believed that this region does not affect the local experiments of relativity and quantum mechanics. However, measurements of the cosmic microwave background radiation from the big bang origin of the universe have established a cosmic preferred frame that provides coupling between local experiments in the macro-world and the external universe that is described in chapter 4.

It is to be stressed that except for the observation of the planets to verify Newtonian mechanics, classical physics was developed in the macro-world. Additionally, the external universe processes cannot be controlled, and the micro-world events with the macro-world instruments are not measurements of parameters because the micro-world parameters are changed by the macro-world events.

1.3   Three Classical Thought Experiments

The thought experiment is an excellent pedagogical device to describe the laws of physics in a way that is consistent with our sense experience–based paradigm for external reality. The advantage over the description of a real experiment for difficult concepts is that unimportant side effects that contribute to errors in real experiments can be eliminated. Both thought experiments and real experiments are used extensively in this book to demonstrate the laws of physics.

Three classical thought experiments that may or may not have been implemented in the real world are analyzed in this section to demonstrate the advantage of this method of describing the laws of physics. The three thought experiments are (1) a demonstration of Archimedes' principle, (2) Galileo's Leaning-Tower-of-Pisa experiments, and (3) Newton's apple experiment.

Archimedes' principle states that a body immersed in a liquid experiences an upward force that is equal to the magnitude of the weight of the liquid it displaces. The legend about Archimedes' discovery is that he realized this principle while stepping into a full bath and was so excited with this discovery that he got out of the bath and rushed naked into the street yelling, "Eureka! Eureka!"

A thought experiment demonstrating Archimedes' principle is shown in figures 1.2a through 1-2c. At the start of the experiment the sailboat is floating in the water where the bottom of the hull is displacing a certain volume of water as shown in figure 1-2a. Then a cast of the volume of the hull that is under water is made with a rigid, lightweight material with zero thickness filled with water as shown in figure 1-2b. Because the water does not move, the weight of the water in each small section of the cast is exactly balanced by the force that the water outside the cast applies to the cast (i.e., any volume of water is stable because its weight is exactly balanced by the force from the water outside the volume). The water is then removed from the cast and replaced with a volume of lead equal in weight to the water (approximately 10%). The force of the water outside the cast will still be the same and exactly support the weight of the lead, which demonstrates that a body immersed in a liquid experiences an upward force that is equal to the magnitude of the weight of the liquid it displaces.

According to legend, Aristotle believed a ten-pound weight would fall ten times faster than a one-pound weight. However, in *Galileo's Leaning-Tower-of-Pisa experiment* he dropped a ten-pound weight and a one-pound weight and they both fell at the same speed. Galileo described the experiment in his writings, but he never claimed to have performed it. It is believed this was only a thought experiment.

This thought experiment is shown in figure 1-3. Here it is assumed that three heavy unconnected balls are dropped from the tower at the same time as shown. By symmetry, they will all fall at the same speed. Then, if ball number 2 and ball number 3 are connected with a lightweight rod as shown by the dotted line in figure 1-3, the three balls will still fall at the same speed because the horizontal rod can't change the vertical speed of the balls.

Now, if ball number 2 and ball number 3 are replaced with a single ball the same size as ball number 1 that is constructed with a material that is ten times the density of ball number 1, then by symmetry the new ball and ball number 1 will fall at the same speed. This demonstrates that Aristotle was wrong, and a ten-pound weight will fall at the same speed as a one-pound weight.

Newton's apple experiment legend is that he was sitting under an apple tree when an apple fell on his head, and he suddenly discovered the universal law of gravitation. A more reasonable version of this legend is that Newton observed an apple fall from a tree and realized that the same force that was causing the apple to fall to the earth was also causing the moon to be circling the earth.

The thought experiment demonstrating Newton's discovery is shown in figure 1-4. Here the gravitational force is the *cause* of the *effect* of the apple falling to the earth as well as the *effect* of the moon falling toward the earth. However, the tangential velocity of the moon in its orbit keeps it from actually falling to the earth. Newton's real discovery was that the force acting on the apple is the same force acting on the moon and represents the universal law of gravitation that applies everywhere in the universe.

These three thought experiments were selected because they represent a nonmathematical description of the concepts that are the basis of classical physics. Archimedes' principle demonstrates the effect of gravitational force (weight) acting on a body close to the earth, as well as the use of logic to establish cause-and-effect relationships in physics. Galileo's leaning-tower experiment demonstrates the use of symmetry to establish the equivalence of gravitational mass (weight) and inertial mass that Einstein used as a starting point for general relativity theory. Finally, Newton's apple experiment demonstrates that gravitational attraction is a universal force acting on local bodies (the apple) as well as distant bodies (the moon).

This chapter establishes the concepts to be used to describe the physical reality defined by the mathematical models of relativity and quantum mechanics. Even though Newton's absolute time must be replaced with three types of time (proper, coordinate, and cosmic) and Newton's absolute space must be replaced with length changing measuring rods, the physical reality of modern physics described in this book is simpler than the conventional interpretation because it is referenced to the observer-independent cosmic preferred frame established by the now-known structure of the universe.

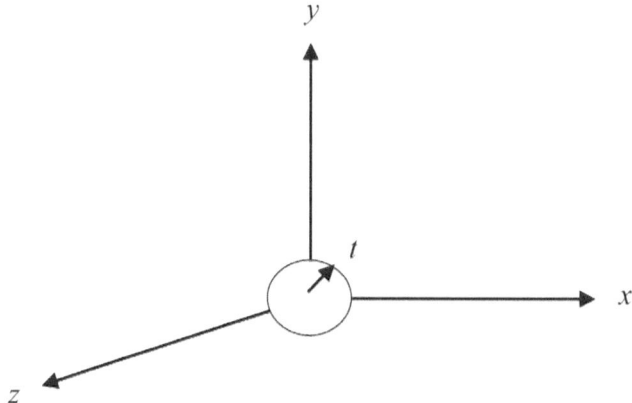

Figure 1-1. A typical classical reference frame

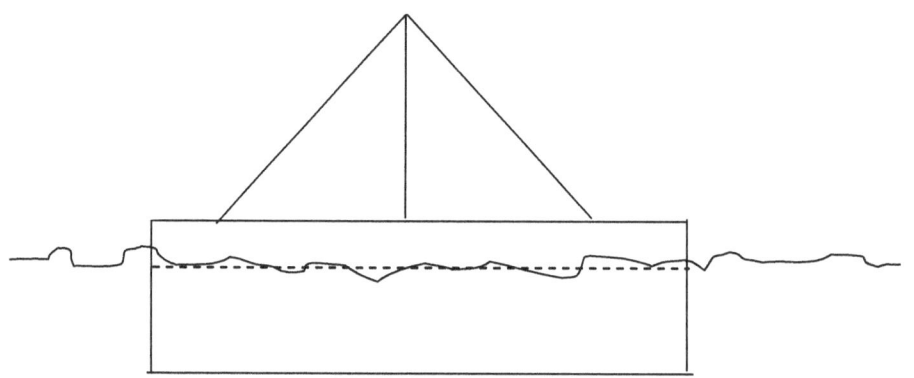

Figure 1-2a. The Archimedes' principle experiment

Figure 1-2b. The hull casting filled with water

Figure 1-2c. The hull casting filled with lead

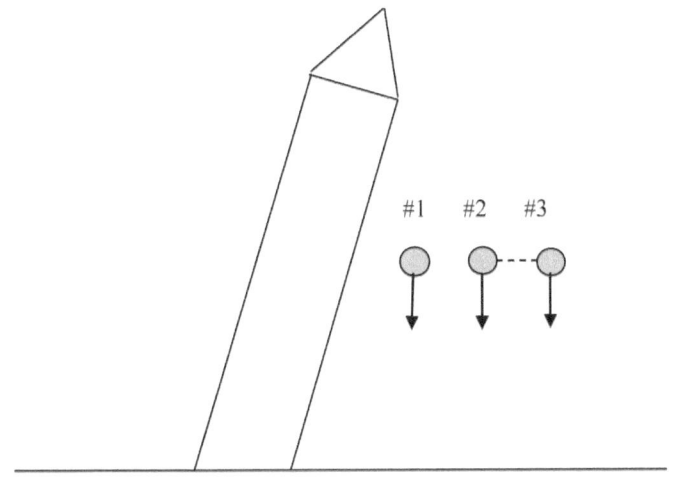

Figure 1-3. Galileo's Tower-of-Pisa experiment

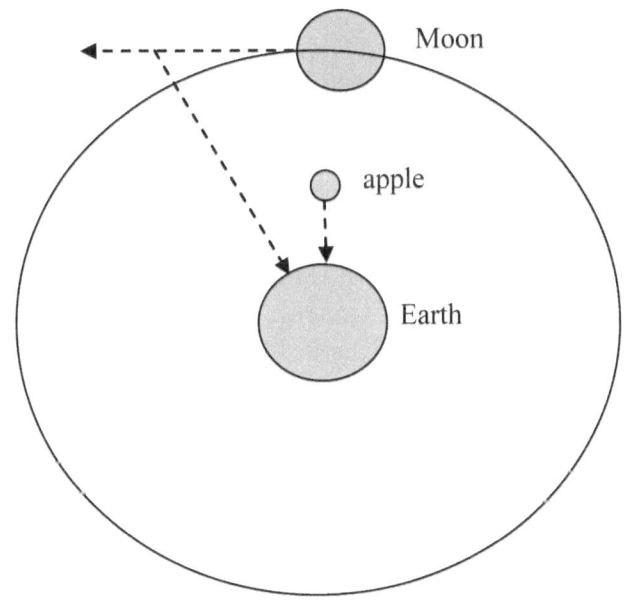

Figure 1-4. Newton's apple experiment

2.0   The Lorentz Preferred Frame (Ether) Theory

At the end of the nineteenth century, primarily because of the form of Maxwell's equations that defined the propagation of electromagnetic waves, it was believed that light was propagated by a medium—the ether—that established a cosmic preferred frame. This preferred frame was consistent with Newton's absolute space and absolute time. Therefore, the physical reality of the Lorentz ether theory was consistent with classical reality.

In 1887, A. A. Michelson and E. W. Morley conducted an experiment to measure the velocity of the earth with respect to the preferred frame of the ether. The null result of this experiment demonstrated that the round-trip speed of light measured on earth is the same in every direction at all times of the year. It will be shown in the following analysis of the experiment that this requires a slight contraction of solid bodies moving through the ether. Because of this, Lorentz and Fitzgerald proposed the contraction of solid bodies along with the time dilation of processes (e.g., clocks) moving through the ether.[2-1]

> The problem of determining the influence exerted on electric and optical phenomena by a translation . . . admits of a comparatively simple solution, so long as only those terms need be taken into account, which are proportional to the first power of the ratio between the velocity of translation v and the velocity of light c. Cases . . . of the order v^2/c^2 . . . present more difficulties. *The first example of this kind is Michelson's well-known interference-experiment, the negative result of which has led Fitzgerald and myself to the conclusion that the dimensions of solid bodies are slightly altered by their motion through the ether.*

2.1   Reference Frames

In the Lorentz ether theory or special relativity theory, events are described with respect to observers located in inertial frames of reference. An inertial frame is defined as a reference frame in which Newton's first law is measured to be true at every point in the frame (i.e., an object that experiences zero external force will move in a straight line at a constant velocity). An observer can be a person, camera, etc., that has the capability of recording events with respect to their locations (i.e., distance and time coordinates) in the rest frame of the observer. The frame of reference can be thought of as an imaginary structure consisting of evenly spaced unit length "rigid" rods, with a clock located at the junction of each of the rods. Each of the clocks is assumed to be synchronized to the master clock located at the origin of the reference frame such that the measured velocity of light in all directions is equal to the constant c. Here an event is represented by four parameters, where the x,y, and z parameters represent the distance to the origin and the t parameter represents the coordinate time of the event.

Rindler has defined the two different descriptions of the external world from an xyzt frame from the point of view of the primary observer located at the origin.[2-2]

> What he actually or potentially sees at any one instant is called his *world-picture* at that instant . . . A much more useful concept is the *world-map*. This, as the name implies, may be thought of as a mapping of events into an observer's instantaneous space t = t_0: a kind of three-dimensional snapshot exposed everywhere simultaneously.

The world-map of the xyzt frame is shown in figure 2-1. Here all of the clocks are synchronized to the origin clock. The world-picture of this situation would show the pointer on each of the clocks earlier by an amount equal to r/c, where r is the distance to the origin and c is the velocity of light.

A thought experiment demonstrating Einstein's method of clock synchronization that ensures that the measured velocity of light will be equal to c in all directions is shown in figure 2-2. Here a pulsed light beam is sent from origin clock a at $t_{a0} = 0$ to the b clock to be synchronized, once each second. When the beam contacts sensor b, the b clock is set to $t_{b0} = 0$. Then, after the reflected beam contacts sensor a at t_{a1}, the value $t_{a1}/2$ is added to clock b using the communication channel.

2.2 The Michelson-Morley Experiment

A simplified version of the interferometer arrangement that was used in the Michelson-Morley experiment is shown in figure 2-3a. Here light from the source is sent to the half-silvered glass plate P placed at forty-five degrees to the beam, which divides the beam into two parts. One part of the beam is sent to mirror M_1 and is then reflected back and transmitted to the target. The other part of the beam is sent to mirror M_2 and is reflected back and transmitted to the target. The distance from plate P to mirror M_1 and plate P to mirror M_2 is equal to L with respect to the xyzt rest frame of the apparatus. Therefore, if the xyzt frame of the interferometer is at rest with respect to the ether frame, by symmetry the time to traverse the horizontal and vertical paths of the apparatus is equal to $2L/c$ with respect to the xyzt frame.

The world-map for the interferometer moving with velocity V_X with respect to the ether frame (XYZT) is shown in figure 2-3b. Because of the null result of the experiment in this situation, the time to travel the vertical path (T_V) and the horizontal path (T_H) represented by the dotted lines must still be equal. The vertical and horizontal path lengths are calculated in appendix 2A and demonstrate that the distance between the half-silvered glass plate P and mirror M_2 (X_L) is equal to $L\sqrt{1 - V_X^2/c^2}$. This is the Lorentz contraction effect of special relativity theory. Lorentz proposed that even though the dimensions of solid bodies appear to be constant to us, in reality these dimensions are modified by changes in the electrical forces acting between the large distances separating the molecules of the solid bodies moving with respect to the ether frame.

In a similar fashion, if the interferometer is considered to be a clock, where each round trip of the light beam is a "tick" of the light clock with respect to the XYZT frame ($T_V = T_H$), then as shown in appendix 2B, the time interval of the "tick" of the clock with respect to the XYZT frame is greater than the "tick" of the clock with respect to the xyzt frame ($2L/c$) by $1/\sqrt{1 - V_X^2/c^2}$. This slowing of the interferometer clock is the time dilation effect of the Lorentz theory or special relativity theory.

2.3   The Lorenz Transformation Equations

Lorentz used the results of this experiment to develop the transformation equations that transform the coordinates of an event (x, y, z, t) with respect to a frame moving with velocity V_x with respect to the ether frame to the coordinates of the ether frame (X, Y, Z, T). The Lorentz transformation equations are given in appendix 2C. The use of these equations to transform the coordinates of events with respect to the xyzt frame to the ether frame (XYZT) is demonstrated by the following analysis of the Michelson-Morley experiment. As shown in figure 2-3b, the four events describing the experiment with respect to the rest frame of the interferometer are

E_0 = The start of the experiment with coordinates $x_0 = 0, y_0 = 0, z_0 = 0, t_0 = 0$;

E_1 = The vertical light beam is coincident with mirror M_1 with coordinates $x_1 = 0, y_1 = L, z_1 = 0, t_1 = L/c$;

E_2 = The horizontal light beam is coincident with mirror M_2 with coordinates $x_2 = L, y_2 = 0, z_2 = 0, t_2 = L/c$; and

E_3 = Both the vertical and horizontal beams are coincident with the half-silvered glass plate P with coordinates $x_3 = 0, y_3 = 0, z_3 = 0, t_3 = 2L/c$.

The value of X_L is calculated in appendix 2C by using the Lorentz transformation equations to transform the x, y, z, t coordinates of event E_2 to the coordinates of the ether frame (XYZT). Again, the value of X_L is equal $L\sqrt{1 - V_x^2/c^2}$, which is the Lorentz contraction effect.

The time required for the velocity of light to travel the round trip vertical path $(2T_1)$ with respect to the XYZT frame is calculated in appendix 2C by using the Lorentz transformation equations to transform the x, y, z, t coordinates of event E_3 to the coordinates of the ether frame (XYZT). Again, the time interval of the "tick" of the clock with respect to the XYZT frame is greater than the "tick" of the clock with respect to the xyzt frame $(2L/c)$ by $1/\sqrt{1 - V_x^2/c^2}$. This slowing of the moving interferometer clock is the time dilation effect of the Lorentz ether theory or special relativity theory.

It is to be stressed that Lorentz assumed these changes in the dimensions of solid bodies and the rate of clocks moving with respect to the ether frame are real. Therefore, the preferred frame of the ether is consistent with Newton's absolute space and absolute time of classical physics. However, because there was no theory to describe why this was true, or any experiment to verify the ether, the Lorentz ether theory was abandoned.

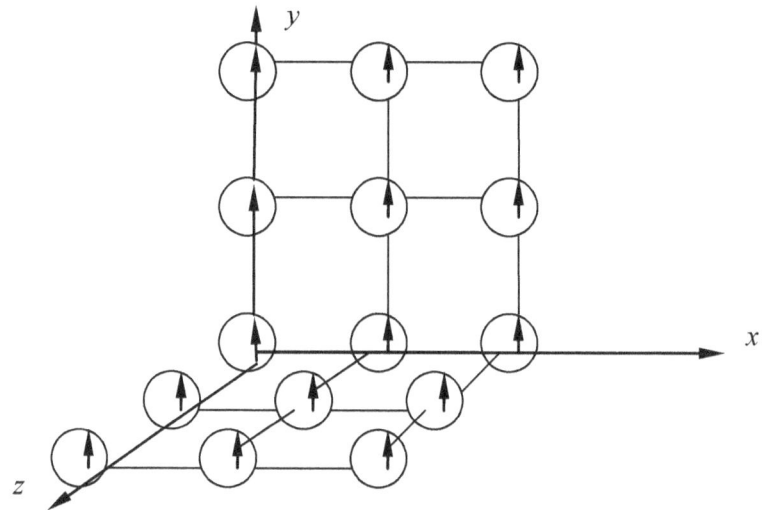

Figure 2-1. The world-map of a reference frame

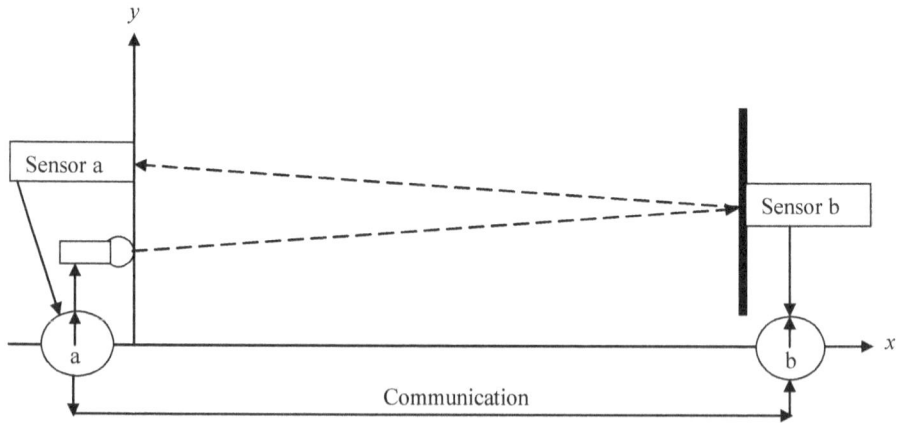

Figure 2-2. Einstein's light beam method of synchronization.

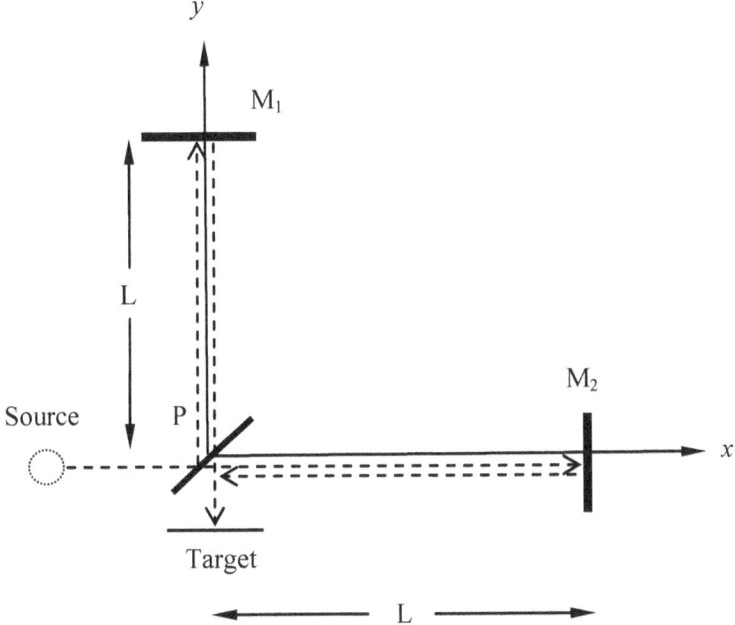

Figure 2-3a. The Michelson-Morley Experiment
referenced to the xyzt frame.

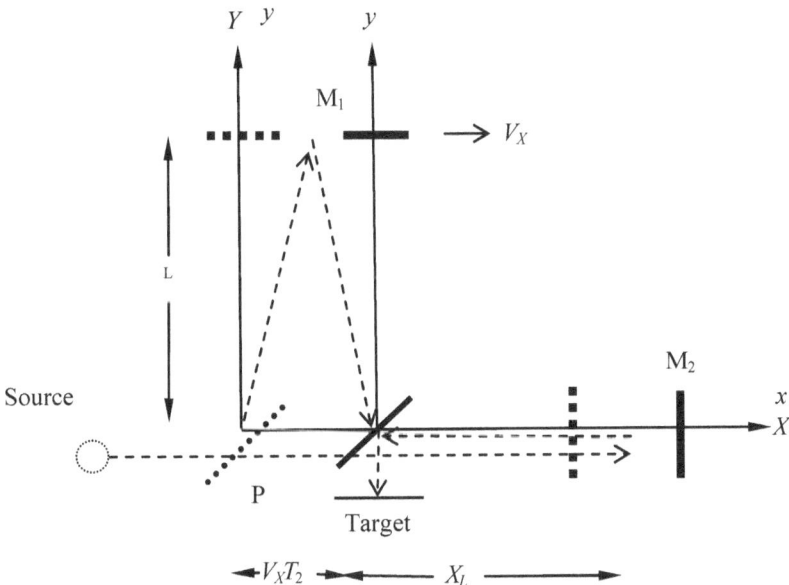

Figure 2-3b. The Michelson-Morley Experiment
referenced to the preferred frame.

3.0   Einstein's Special Relativity Theory

Einstein's special relativity theory is difficult to comprehend because there are no cause-and-effect relationships to explain the physical reality of slower clocks (time dilation) and shorter measuring rods (length contraction) for moving frames in an otherwise empty universe. Even though there was no experimental verification of the existence of the ether, the Lorentz preferred frame (ether) theory was consistent with our sense-based paradigm for physical reality because the *cause* of the *effects* of time dilation and length contraction is the motion of the clocks and measuring rods with respect to the cosmic preferred frame.

In his 1905 paper that established special relativity theory, Einstein took a totally different approach than Lorentz and assumed there was no need to postulate an ether that was untestable at that time.[3-1]

> Examples of this sort, together with the unsuccessful attempts to discover any motion of the earth relatively to the "light medium," suggest that the phenomena of electrodynamics as well as of mechanics possess no properties corresponding to the idea of absolute rest. They suggest rather that, as has already been shown to the first order of small quantities, the same laws of electrodynamics and optics will be valid for all frames of reference for which the equations of mechanics hold good . . . The introduction of a "luminiferous ether" will prove to be superfluous inasmuch as the view here to be developed will not require an "absolutely stationary space" provided with special properties, nor assign a velocity vector to a point of the empty space in which electromagnetic processes take place.

While there have been questions raised about Einstein's knowledge of the Michelson-Morley experiment at the time of the paper, in a letter to Jaffe, Einstein stated,[3-2]

> It is no doubt that Michelson's experiment was of considerable influence upon my work insofar as it strengthened my conviction concerning the validity of the principle of the special theory of relativity. On the other side I was pretty much convinced of the principle before I did know this experiment and its results.

He formally defined the principle of relativity in his 1905 paper with two postulates:[3-3]

1. The laws by which the states of physical systems undergo change are not affected, whether these changes of state are referred to the one or the other of two systems of coordinates in uniform translatory motion.

2. Any ray of light moves in the "stationary" system of coordinates with the determined velocity c, whether the ray be emitted by a stationary or a moving body. Hence

$$\text{velocity} = \frac{\text{light path}}{\text{time interval}}$$

Einstein then used these two postulates to derive the equations to transform the coordinates of an event referenced to one inertial frame to the coordinates of another inertial frame moving with respect to it. These transformation equations are the Lorentz transformation equations that were derived by Lorentz to describe the results of the Michelson-Morley experiment referenced to the ether frame. Unfortunately, because the postulates of special relativity theory are not consistent with our sense experience–based concept of external reality, the results of applying the Lorentz transformation equations to the special relativity experiments are not consistent with our intuitive concepts of space and time (e.g., observers in each of two inertial frames moving with respect to each other will measure the same value for the speed of light).

M. S. Longair has accurately stated the problem:[3-4]

> Let me state immediately that there are no paradoxes—only some
> rather non-intuitive features of space-time which result from the
> Lorentz transformations . . . . I am very suspicious of arguments
> which try to give you simple ways of looking at relativistic
> problems. Things like length contraction and time dilation should
> be treated cautiously and the simplest way not to get horribly
> confused is to write down the four vectors associated with the events
> you are considering and then use the Lorentz transforms to relate
> coordinates.

3.1   The Analysis of Special Relativity Experiments

The conventional special relativity analysis of clock experiment 1
given in the next section demonstrates the time dilation and Lorentz
contraction effects for two inertial frames moving with velocity v with
respect to each other. It is not necessary to understand the mathematical
analysis of this experiment to realize that figures 3-1a, 3-1b, and 3-1c give
a commonsense explanation for these effects based on the definitions for
proper time and *coordinate time*. Here proper time is defined as the time
measured by a single clock coincident with each of two events, while
coordinate time is defined as the time between two events measured by
the difference in the proper time of two synchronized clocks in a frame,
each coincident with one of the events. This experiment demonstrates
that the proper time measured by each frame's origin clock is less than
the coordinate time of the other frame by a factor of $\sqrt{1 - v^2/c^2}$.

The analysis of portable clock frame synchronization is given in
section 3.3. It is not necessary to understand the mathematics of the
analysis to realize that figure 3-2b demonstrates that the time dilation
effect of a slowly transported portable clock results in the proper time
of the clock matching the proper time of any synchronized clock in the
frame. While the rest of this chapter using figures 3-1a, 3-1b, and 3-1c,
and appendixes 3A, 3B, and 3C is important to verify the definitions
of proper time and coordinate time, along with synchronization using
a slowly transported portable clock, it can be skipped because Bell's
preferred frame interpretation of special relativity is based on the
preferred frame Lorentz ether theory described in the last chapter.

3.2   Clock Experiment 1

The purpose of clock experiment 1, shown in figures 3-1a and 3-1b, is to demonstrate the special relativity theory effects of time dilation and the Lorentz contraction for two inertial frames moving with respect to each other. Here it is assumed that there are three clocks (a, b, and c) that are firmly attached to metal rods that are at rest with respect to the xyzt frame. The clocks are synchronized with Einstein's light beam method of synchronization, and the b clock is located at the origin of the xyzt frame.

In a similar fashion, there are three clocks (a', b', and c') that are firmly attached to metal rods that are at rest with respect to the x'y'z't' inertial frame that is moving with velocity v_x with respect to the xyzt frame. The clocks are synchronized with Einstein's light beam method of synchronization, and the b' clock is located at the origin of the x'y'z't' frame. Additionally, it is assumed that the b and b' clocks each display zero when the origins of the two frames are coincident. This ensures that the time of the a, b, c, and b' clocks all display zero as shown in figure 3-1a.

The location and time displayed by the a' and c' clocks with respect to the local reference frame (xyzt) can be calculated using the Lorentz transformation equations that are used to transform the space (x', y', and z') and time (t') coordinates of events from the moving x'y'z't' frame to the xyzt reference frame.

The events defining the location and time displayed by the a' and c' clocks at $t_0 = 0$ are

$E_{a'0}$ = The a' clock is located at $x'_{a'0} = -L$ and $t_0 = 0$

$E_{c'0}$ = The c' clock is located at $x'_{c'0} = L$ and $t_0 = 0$

The Lorenz transformation equations are used in appendix 3A to calculate the time displayed (i.e., coordinate time of the x'y'z't' frame) and the space coordinates of the a' and c' clocks with respect to the xyzt frame at $t_0 = 0$.

$$x_{a'0} = -L\sqrt{1 - v_x^2/c^2} \qquad\qquad t'_{a'0} = Lv_x/c^2$$

$$x_{c'0} = L\sqrt{1 - v_x^2/c^2} \qquad\qquad t'_{c'0} = -Lv_x/c^2$$

The world-map of this configuration is shown in figure 3-1a. The location of the a' and c' clocks demonstrate the Lorentz contraction effect of special relativity theory (i.e., each of the metal rods supporting the clocks is contracted by $\sqrt{1 - v_x^2/c^2}$).

In order to demonstrate the special relativity time dilation effect, the experiment at time $t_1 = (L/v_x) \sqrt{1 - v_x^2/c^2}$ is shown in figure 3-1b. Here clock a' has moved to the origin of the xyzt frame as shown. The events defining the location and time displayed by the a' and b' clocks at $t_1 = (L/v_x) \sqrt{1 - v_x^2/c^2}$ are,

$$E_{a'1} = \text{The a' clock is located at } x'_{a'1} = -L \text{ and } t_1 = (L/v_x) \sqrt{1 - v_x^2/c^2}$$

$$E_{b'1} = \text{The b' clock is located at } x'_{b'1} = 0 \text{ and } t_1 = (L/v_x) \sqrt{1 - v_x^2/c^2}$$

The Lorentz transformation equations are used in appendix 3A to solve for the time displayed (i.e., coordinate time of the x'y'z't' frame) and the space coordinates of the a' and b' clocks with respect to the xyzt frame at t_1.

$$x_{a'1} = 0 \qquad\qquad t'_{a'1} = L/v_x$$

$$x_{b'1} = L\sqrt{1 - v_x^2/c^2} \qquad\qquad t'_{b'1} = (L/v_x)(1 - v_x^2/c^2)$$

The world-map of this configuration is shown in figure 3-2b.

Figures 3-1a and 3-1b demonstrate the two different types of time that exist because of the different synchronization of the x'y'z't' and xyzt frames. *Proper time* is defined as the time measured by a single clock coincident with each of two events, while *coordinate time* is defined as the time between two events measured by the difference in the proper time of two synchronized clocks in a frame, each coincident with one of the events. For example, the time dilation of the proper time of the b' clock ($t'_{b'1} - t'_{b'0}$) is measured with respect to the coordinate time of the xyzt frame ($t_1 - t_0$) that is the difference between the proper time of the b clock at $t_0 = 0$ and the proper time of a synchronized clock in the xyzt frame coincident with the b' clock at $t_1 = L/v_x \sqrt{1 - v_x^2/c^2}$. This difference between the proper time of the b' clock and the coordinate time of the xyzt frame is calculated in appendix 3A. The proper time of the b' clock is less than the coordinate time of the xyzt frame by a factor of $\sqrt{1 - v_x^2/c^2}$.

The difference between the proper time of the b clock ($t_{b1} - t_{b0}$) and the coordinate time of the x'y'z't' frame ($t'_{a'1} - t'_{b'0}$) is calculated in appendix 3A. The proper time of the b' clock is less than the coordinate time of the x'y'z't' frame by a factor of $\sqrt{1 - v_x^2/c^2}$.

These equations show there is no inconsistency in the fact that an observer at the origin of the reference frame (xyzt) will determine that the clocks located in the moving x'y'z't' frame run at a slower rate than his or her coordinate time, while an observer at the origin of the moving x'y'z't' frame will determine that the clocks in the reference frame (xyzt) run slower than his or her coordinate time. This is caused by the difference in the synchronization of the two frames.

Figure 3-1c shows the beginning of this experiment referenced to the synchronization of the x'y'z't' frame at $t'_0 = 0$. The events defining the location and time displayed by the a and c clocks are

E_{a0} = The a clock is located at $x_{a0} = -L$ and $t'_0 = 0$

E_{c0} = The c clock is located at $x_{c0} = L$ and $t'_0 = 0$

The Lorentz transformation equations are used in appendix 3B to solve for the displayed time and coordinates of the a and c clocks with respect to the x'y'z't' frame at $t'_0 = 0$.

$$x'_{a0} = -L\sqrt{1 - v_x^2/c^2} \qquad\qquad t_{a0} = -Lv_x/c^2$$

$$x'_{c0} = L\sqrt{1 - v_x^2/c^2} \qquad\qquad t_{c0} = Lv_x/c^2$$

The world-map of this configuration is shown in figure 3-1c. This demonstrates that the analysis of clock experiment 1 is exactly the same for an observer at the origin of the xyzt frame or an observer at the origin of the x'y'z't'. The description of clock experiment 1 given by figures 3-2a, 3-2b, and 3-2c demonstrates the advantage of using the definitions for proper time and coordinate time to describe the special relativity effects of time dilation and the Lorentz contraction.

3.3   Portable Clock Frame Synchronization

The purpose of this experiment is to demonstrate that clock synchronization using a slowly transported portable clock is equivalent to clock synchronization using Einstein's method of light beam synchronization. The approach is shown in figure 3-2a. Here it is assumed that there are two clocks (a' and b') that have been synchronized with Einstein's method of light beam synchronization that are attached to a metal rod of length L that is at rest with respect to the x'y'z't' frame. In order to verify that portable clock p can be used to synchronize clock b', clock p is slowly transported from clock a' to clock b' with velocity v'_p, and the proper time of clock p is compared to clock b'.

The two events defining the experiment with respect to the x'y'z't' frame are

$E_{a'0}$ = Start of the experiment where clock p is set to 0 and is coincident with clock a' at $x'_{a'0} = 0$ and x'y'z't' coordinate time $t'_0 = 0$, and

$E_{b'1}$ = End of the experiment where clock p is coincident with clock b' at $x'_{b'1} = L$ and x'y'z't' coordinate time $t'_1 = L/v'_p$.

The proper time of clock p at event $E_{b'1}$ (t_{p1}), calculated in appendix 3C, is equal to $t'_{b'1}$ as shown in figure 3-2a.

Figure 3-2b is a world-map of the portable clock synchronization with respect to the x'y'z't' frame, where the x'y'z't' frame is moving with velocity v_x with respect to the xyzt reference frame. It is not necessary to follow the mathematics of this analysis in appendix 3C to understand that this experiment demonstrates that if a portable clock (p) is moved to a synchronized clock (a') in an inertial frame and set to the proper time (i.e., coordinate time $t'_{a0} = 0$) of the synchronized clock, and then if the portable clock is slowly moved to a second synchronized clock (b') in the inertial frame, the proper time of the portable clock (t_{p1}) will be identical to the proper time of the second synchronized clock $(t'_{b'1})$ in the inertial frame. This property of a slowly transported portable clock can be used to synchronize the clocks in an inertial frame instead of the usual method of Einstein's light beam synchronization.

The examples of time dilation and length contraction based on proper time and coordinate time given in this chapter demonstrate that the mathematical models of special relativity theory are correct for local experiments, but do not provide the cause-and-effect relationships necessary to establish the physical reality of these models.

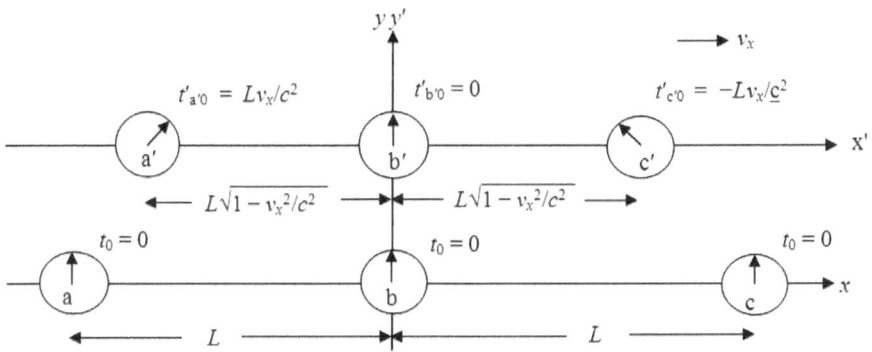

Figure 3-1a. Clock experiment 1 referenced
to the xyzt frame at t_0

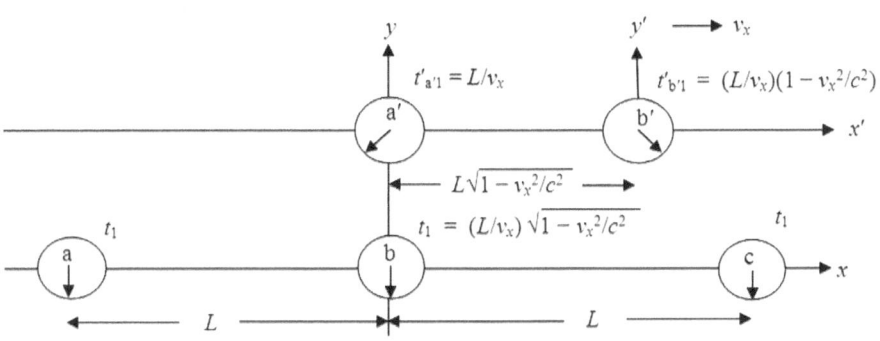

Figure 3-1b. Clock experiment 1 referenced
to the xyzt frame at t_1

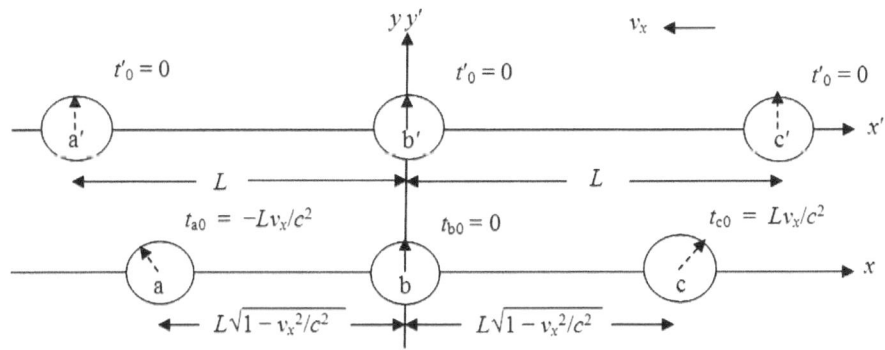

Figure 3-1c. Clock experiment 1 referenced
to the x'y'z't' frame at t'_0

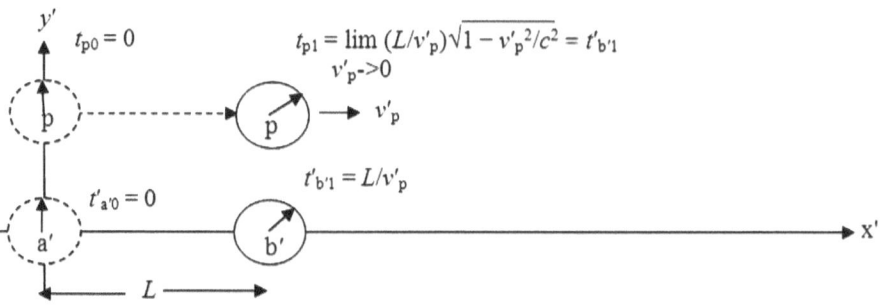

Figure 3-2a. Portable clock synchronization of
the x'y'z't' frame at coordinate time t'_1

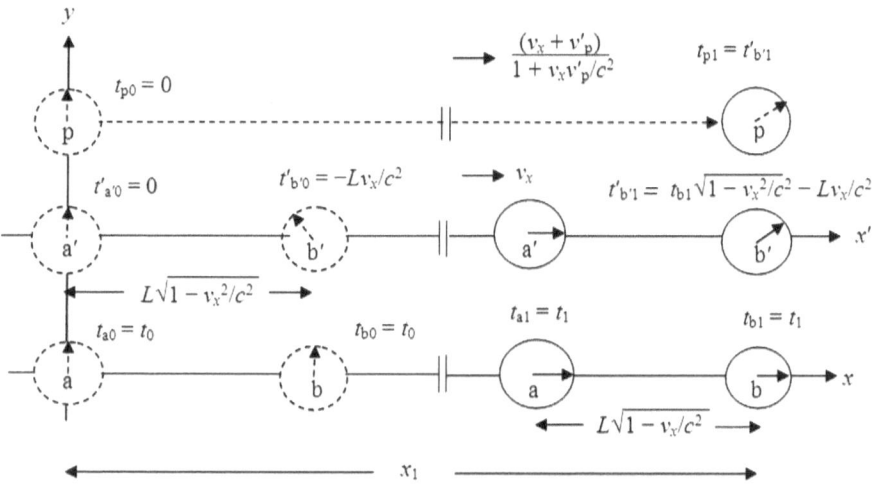

Figure 3-2b. Portable clock synchronization of
the x'y'z't' frame at coordinate time t_1

4.0   An Extension of Bell's Special Relativity Theory

John Bell was one of the few physicists after Einstein who was concerned with the reality behind the mathematical models of modern physics. He believed that an observer-independent preferred frame description of special relativity is much more consistent with our sense experience–based perception of external reality, as well as being in agreement with all of the special relativity experiments. However, his work was basically ignored during his lifetime because there was no experimental support for a cosmic preferred frame.

Using the now-known observer-independent preferred frame structure of the universe and postulating the equivalence of special relativity time dilation and gravitational time dilation, an extension of Bell's preferred frame interpretation is developed in this chapter that restores the *cause*-and-*effect* relationship to the mathematics of special relativity theory, eliminates the conventional belief that time travel to the past is possible, and verifies that the centrifugal force in a rotating frame results from the motion with respect to the mass/energy in the external universe, thus eliminating the conventional belief in the possibility of an "empty" universe. The advantage of this interpretation of relativity theory is then demonstrated by the analysis of five important relativity experiments.

4.1   Bell's Interpretation of Special Relativity Theory

In a BBC interview Bell stated his belief in a preferred frame (ether) interpretation of special relativity theory.[4-1]

> Bell: "*Well, what is not sufficiently emphasized in textbooks, in my opinion, is that the pre-Einstein position of Lorentz and Poincare, Larmor, and Fitzgerald was perfectly coherent, and is not inconsistent with relativity theory.* The idea that there is an ether, and these Fitzgerald contractions and Larmor dilations occur, and that as a result the instruments do not detect motion through the aether—that is a perfectly coherent point of view."
> Narrator: "And it was abandoned on grounds of elegance?"
>
> Bell: "Well, on the grounds of philosophy; that what is unobservable does not exist. And also on grounds of simplicity, because Einstein found that the theory was both more elegant and simpler when we left out the idea of the aether. *I think that the idea of the ether should be taught to students as a pedagogical device, because I find that there are lots of problems, which are solved more easily by imagining the existence of an ether.* But that's another story."

However, Marder accurately states that the conventional interpretation of special relativity theory assumes there is no coupling between local experiments and the external universe (ether).[4-2]

> Special relativity does not therefore deny the existence of the background matter of the universe. It merely regards whatever role that matter might have in determining the inertial motions as outside its province, and takes the observed existence of these preferred motions as a starting point.

This conventional interpretation of special relativity theory results in many *cause*-and-*effect* problems in interpreting special relativity experiments. For example, if a portable clock is accelerated and travels between two points on the earth at sea level, then because of the time dilation effect the moving clock will run slower than synchronized clocks on the earth as a function of its velocity with respect to the earth (i.e., if the velocity of the clock is much less than the velocity of light, the time measured by the moving clock will almost be the same as the earth clocks, while if the velocity of the moving clock is close to the speed of light, the time measured by the clock will be close to zero with respect to the earth clocks). The real paradox of special relativity theory time dilation is that if it is assumed that there is no coupling between the moving clock and objects located on the earth, and the rest of the universe can be ignored, *then clocks with identical environments run at different rates* (i.e., no cause-and-effect relationship to explain special relativity time dilation). Bell's answer to this time dilation paradox is that the special relativity effects of time dilation and the Lorentz contraction are caused by motion with respect to the cosmic preferred (ether) frame that changes the environment of the portable clock as a function of its velocity with respect to the ether.

The advantage of this preferred frame interpretation of special relativity theory is demonstrated by clock experiment 2 shown in figures 4-1a and 4-1b. Here the xyzt frame is moving with velocity V_X with respect to the XYZT ether frame. Clocks a and b are located in the xyzt frame a distance x_1 apart, and clocks A and B are located in the XYZT frame a distance X_1 apart as shown. The events defining the experiment are

E_0 = The origin of the xyzt frame, the origin of the XYZT frame, and the portable p clock are all coincident at $t_a = 0$, $T_A = 0$, and $t_p = 0$ at the start of the experiment.

E_1 = The p clock has moved to x_1, y_1, and t_1 with respect to the xyzt frame at the end of the experiment.

With respect to figure 4-1a, the time dilation effect is used to calculate the proper time of clock p (t_{p1}) with respect to the xyzt frame at the end of the experiment in appendix 4A. Then, the Lorentz transformation equations are used to calculate the proper time of clock p (t_{p1}) referenced to the XYZT frame at the end of the experiment. The form of these two equations demonstrates that the time dilation effect of the proper time of a clock moving in an inertial frame that is moving with respect to the cosmic preferred frame is caused by the velocity of the clock referenced to the cosmic preferred frame, independent of the velocity of the inertial frame with respect to the cosmic preferred frame.

The pedagogical advantages of John Bell's preferred frame interpretation are the following:

1. The cosmic preferred frame restores Newton's absolute time and absolute space for the external universe and restores the classical past, present (NOW), and future. This eliminates the possibility of time travel to the past because the past no longer exists.
2. The local special relativity *effects* of time dilation and the Lorentz contraction are *caused* by motion with respect to the cosmic preferred frame.
3. Bell pointed out that the preferred frame interpretation of special relativity theory is the "cheapest" resolution of the inconsistency between Aspect's experimental violation of Bell's inequality and the conventional interpretation of special relativity theory. This will be described in chapter 6.

Unfortunately, because of Einstein's legacy and the inability to establish the presence of the ether in the past, the conventional belief is that there can be no cosmic preferred frame. Lee Smolin has accurately stated this position.[4-3]

> *I have a lot of trouble believing that special relativity is false; if it is, then there is a preferred state of rest and both the direction and speed of motion must be ultimately detectable.*

However, Bell's interpretation of special relativity theory is now supported by measurements of the cosmic microwave background radiation from the big bang origin of the universe that have established a cosmic preferred frame, along with the ability to measure "both the direction and speed of motion" of the earth with respect to this frame.

4.2   The Preferred Frame Structure of the Universe

One of the most important scientific accomplishments in the twentieth century was to establish the structure and age of the universe. First, Hubble verified that the Andromeda Nebula was a galaxy in 1923, along with the realization that the universe was expanding. Then, the conflict between the "steady state universe" and the big bang origin of the universe was resolved by measurements of the cosmic microwave background radiation. These data along with the astronomical observations from observatories on the earth and the Hubble space telescope have given us a very accurate description of the structure of the universe.[4-4]

> *Our universe is dotted with over 100 billion galaxies, and each one contains roughly 100 billion stars.* It is unclear how many planets are orbiting these stars, but it is certain that at least one of them has evolved life. In particular, there is a life form that has had the capacity and audacity to speculate about the origin of this vast universe . . . *The WMAP team estimated that the universe is 13.7 billion years old to within an error of just 0.2 billion years.*

It is very difficult to imagine the number of stars in one hundred billion (10^{11}) galaxies, each containing one hundred billion (10^{11}) stars. However, the following thought experiment can be used to help comprehend these many stars. Assume a box that is two by two by two inches and holds fifty-six Chinese checker marbles. If we imagine that each marble represents one of the stars in our Milky Way galaxy, then one hundred billion marbles would require a volume $200 \times 200 \times 200$ feet. This would be twenty stories high, two-thirds of a football field long and two-thirds of the length of a football field wide. The total number of stars in one hundred billion galaxies each containing one hundred billion stars is represented by a volume of marbles twenty stories high, and thirteen thousand miles long and thirteen thousand miles wide. Thirteen thousand miles is about half the distance around the earth.

In his book *The Very First Light*, John Mather describes the development of the cosmic microwave background measurements from the first detection in 1964 to the COBE satellite that was launched in 1989 that resulted in Nobel prizes for Robert Wilson and Arno Penzias and later for George Smoot and John Mather. These measurements established the existence of the cosmic preferred frame.[4-5]

> Although there had been earlier important antecedents, COBE's story really began in 1964. That year Robert W. Wilson and Arno A. Penzias of Bell Laboratories, while trying to eliminate microwave noise from an antenna designed to pick up satellite signals, discovered a phenomenon known as the cosmic background radiation. The radiation suffused the sky in every direction at microwave

frequencies, and came to be widely regarded as the afterglow of the Big Bang. In the aftermath of this discovery, cosmology became an empirical science. During the 1970s and 1980s, more sophisticated equipment refined the measurements of the cosmic background at different frequencies.

In his book *Wrinkles in Time*, George Smoot points out that the cosmic microwave background radiation not only establishes a cosmic preferred frame, but also the Doppler effect can be used to measure both the direction and speed of motion of the earth with respect to the cosmic preferred frame.[4-6]

> However, all was not as expected. "Look at that," I said to Jon. "What do you suppose that means?" Although the anisotropy was close to the magnitude we had expected, its direction was nearly the opposite. That is, the sky was warmest in the direction of Leo and coolest in the direction of Aquarius, which means that Earth was moving toward the former and away from the latter. That is *not* the direction in which the Galaxy rotates. "Unless we have a sign wrong," said Jon, "there's only one explanation." We both knew what the answer had to be: *Not only is the entire Galaxy rotating, as it should be, but, unexpectedly, it is also moving through space. And it was moving very fast—six hundred kilometers a second, or more than a million miles an hour.*

Additionally, Brian Greene describes the *cosmic time* of the external universe that can be used to unambiguously define the same time scale from anywhere in the universe.[4-7]

> ... *The uniformity of cosmic evolution throughout all of space allows a physicist in the Milky Way galaxy, and one in the Andromeda galaxy, and another in the Tadpole galaxy to all agree on the universe's history and age.* Concretely, the homogeneous evolution of the universe means that a clock here, a clock in the Andromeda galaxy, and a clock in the Tadpole galaxy will, on average, have been subject to nearly identical physical conditions and hence will have ticked off time in nearly the same way. The homogeneity of space thus provides a universal synchrony.

The existence of the cosmic preferred frame with coordinate time equal to cosmic time is in conflict with the conventional way of teaching special relativity theory. Most special relativity theory textbooks not only deny any coupling between local experiments and the external universe, but also deny the possibility of a preferred frame even though experiments conducted on the cosmic microwave background radiation from the "big bang" origin of the universe over the last forty years have established a cosmic preferred frame.[4-8]

> The principle of relativity specifically states that laboratories moving with a constant velocity are physically equivalent to a laboratory at rest. Therefore, there can be no physical basis for distinguishing a laboratory at rest from another moving at a constant velocity. Imagine that you and I are in spaceships coasting at a constant velocity in deep space. You will consider yourself to be at rest, while I am moving by you at a constant velocity. I, on the other hand, will consider myself to be at rest, while *you* are moving by *me* at a constant velocity. According to the principle of relativity, there is no physical experiment that can resolve our argument about who is at rest. Which of us we consider to be at rest is arbitrary.

While this statement of the principle of relativity is true for local experiments, it is obviously incorrect for the universe as a whole (i.e., the direction and speed of motion with respect to *rest* (the cosmic preferred frame) can be calculated from measurements of the cosmic microwave background radiation using the Doppler effect for each of the spaceships).

4.3   Gravitational and Special Relativity Time Dilation

Even though the relativistic effects of special relativity time dilation and gravitational time dilation change the proper time of a clock moving with respect to a reference frame or located by a gravitational body, respectively, they are treated as totally different effects.[4-9]

> *It is worth noting right from the start of this discussion that this gravitational time dilation is significantly different from the time dilation of the special theory of relativity.* The gravitational effect is

not symmetric in the way that time dilation between inertial reference frames is. Between two inertial reference frames, each measures the other's time as going more slowly than their own. In the general relativistic effect, on the other hand, an observer far from any source of gravity will measure the time of someone near the source as going more slowly than his own, and the observer near the source will measure the distant time as going *faster*. This asymmetry will be shown in the proof of the effect, just as the symmetry in the special relativistic case is revealed in its proof.

In 1959 R. V. Pound and G. A. Rebka verified the gravitational time dilation effect of general relativity in an experiment conducted in a seventy-four-foot tower located in the Lyman Laboratory at Harvard University. This experiment used the Mössbauer effect to measure the redshift of gamma rays. Here there is no paradox as before because the slowing of the clocks is caused by the change in the gravitational potential resulting from the distance to the local gravitational body. The gravitational time dilation effect of general relativity for a clock located a distance R from a mass M is defined in appendix 4B.

Gott gives an interesting example of gravitational time dilation.[4-10]

First, disassemble the planet Jupiter and use its material to construct around yourself an incredibly dense spherical shell whose diameter is just a bit larger than the critical diameter needed for that mass to collapse to a black hole (for a Jupiter-mass shell, that is a bit bigger than 5.64 meters, roomy enough for you to sit inside). *Interestingly, Newton showed that a spherical shell of matter would exert no gravitational effects inside, a result that happens to be true in Einstein's theory of gravity as well.* The bits of mass in the shell would completely surround you and the forces they exert on you would act in all different directions, canceling each other out exactly, leaving a zero net effect. So even though the spherical shell is quite massive, once inside no gravitational forces would affect you . . . *The traveler and the distant observers would both agree that he was aging 4 times more slowly than observers far outside.* As pointed out by astronomer Thomas Gold of Cornell, the time traveler and distant observers outside age differently because their

situations are not symmetrical: the time traveler is deep within a
gravitational well, while they are not.

The fact that the *effect* of gravitational time dilation is *caused* by the
coupling to local gravitational bodies suggests that the *effect* of special
relativity time dilation is *caused* by the coupling to the total mass/energy
in the external universe.

If the *proper time* of a clock or process (e.g., biological aging, etc.)
is defined as the time measured in the rest frame of the clock or process,
then the proper time measured by a clock located close to a local
gravitational body will be less than *cosmic time* because of *gravitational
time dilation*, while the proper time measured by a clock moving with
respect to the cosmic preferred frame will be less than cosmic time
because of *special relativity time dilation*. Because of the mass increase
for moving bodies and the Lorentz contraction effect of special relativity,
it is to be expected that the gravitational potential for a frame moving
with respect to the cosmic preferred frame will be decreased (i.e., the
mass of the distant matter in the universe is greater and is effectively
closer in the direction parallel to the velocity). Therefore, because the
gravitational time dilation effect of general relativity predicts that a
clock runs slower when it is closer to local gravitational bodies (i.e.,
reduced potential), it is postulated that the reason moving clocks run
slower is because of this mass increase and Lorentz contraction-reduced
potential. Referenced to the cosmic time of the cosmic preferred frame,
gravitational time dilation and special relativity time dilation are the same
effect, where gravitational time dilation is caused by the gravitational
potential of local mass and special relativity time dilation is caused by
the change in gravitational potential resulting from the change in the
effective distribution of all of the mass/energy in the external universe
for moving frames.

Here Bell's ether has been replaced by the gravitational field
resulting from all of the mass/energy in the external universe. While
Einstein changed his opinion of this interpretation of spacetime in his
later years, this model is consistent with his views in the early 1920s,
which eliminated the possibility of an empty universe.[4-11]

There can be no space nor any part of space without gravitational
potentials; for these confer upon space its metrical qualities, without

which it cannot be imagined at all. The existence of the gravitational
field is inseparably bound up with the existence of space.

Sciama pointed out that many of the results of relativistic cosmology
are equivalent to the Newtonian dynamics of a large gas cloud.[4-12]

> To prepare the way towards understanding relativistic cosmology
> we shall consider the Newtonian dynamics of a large gas cloud. Not
> only is the Newtonian theory mathematically simpler, it also leads
> to many results that are essentially the same as in relativity, as was
> discovered in 1934 by E. A. Milne and W. H. McCrea.

Therefore, an approximate classical model of the universe can be
constructed by assuming (1) the average mass of the stars in the universe
is equal to the mass of the sun, (2) the Milky Way galaxy has one hundred
billion stars and is one hundred thousand light-years in diameter, (3) the
mass of the Andromeda galaxy is approximately the same as the Milky
Way, and (4) the total universe consists of one hundred billion galaxies
in a spherical volume of radius thirteen billion light-years. An estimate
of the density of the universe for this model is calculated in appendix
4B. The value is 2.8×10^{-30} g·cm^{-3}, which is in reasonable agreement
with modern estimates ($\approx 1.0 \times 10^{-30}$ g.cm^{-3}). Additionally, an equation
is derived that links gravitational time dilation and special relativity time
dilation to the cosmic time of the cosmic preferred frame. Lee Smolin
states the importance of recognizing that special relativity time dilation
and gravitational time dilation are the same effect.[4-13]

> The most CHERISHED goal in physics, as in bad romance novels,
> is unification. *To bring together two things previously understood as
> different and recognize them as aspects of a single entity—when we
> can do it—is the biggest thrill in science.*

The following five experiments demonstrate the advantage of this
cosmic preferred frame interpretation of relativity reality.

4.4   The Hafele-Keating Experiment[4-14]

This experiment is an excellent demonstration that special relativity
time dilation and gravitational time dilation are the same effect. In 1971,

J. C. Hafele and R. Keating verified special relativity time dilation and gravitational time dilation by flying four cesium beam clocks first eastward and then westward around the world on commercial jet airplanes. This experiment is based on the fact that with respect to the earth-centered non-rotating inertial frame above the north pole, for the eastward journey the velocity of the flying clocks is greater than the velocity of the reference clock at rest with respect to the naval observatory in Washington, D.C. Similarly, the velocity of the flying clocks is less than the velocity of the reference clock on the westward journey. Therefore, with respect to the reference clock at the naval observatory, the flying clocks will measure less elapsed time for the eastward journey than for the westward journey. Additionally, the flying clocks will measure greater elapsed time than the reference clock at the naval observatory as a function of altitude because of the gravitational time dilation effect.

The predicted elapsed time differences between the flying clocks and the naval observatory clock were calculated using one hundred twenty-five intervals on the eastward trip and one hundred eight intervals on the westward trip. The special relativity theory time dilation and gravitational time dilation values for the flying clocks were calculated for each interval using the altitude and ground speed of the aircraft, and then added together to get the predicted elapsed time. The measured elapsed time for the naval observatory clock was calculated using the velocity of the earth's surface at Washington D.C. with respect to the earth-centered nonrotating inertial frame. The predicted time differences are shown in table 4-1 and the observed results of this experiment are given in table 4-2.

The preferred frame interpretation of this experiment is consistent with our sense experience–based interpretation of external reality because it demonstrates that special relativity time dilation and gravitational time dilation are the same effect, where gravitational time dilation results from the distribution of local mass/energy and SRT time dilation is caused by the change in the effective distribution of all of the mass/energy in the external universe for a moving frame.

4.5   The Special Relativity Twin Paradox Revisited

Probably the best known example of the difficulty of reconciling our sense experience–based concept of time with the time dilation effect of special relativity theory is given by the infamous twin experiment.

Here it is assumed that one twin pilots a rocket ship to a nearby star and back by (1) using the rocket engines to accelerate from the earth in the direction of the star, (2) coasting to the star, (3) using the rocket engines to accelerate in the direction of earth, (4) coasting back to earth, and (5) applying the rocket engines to land on earth. If the acceleration times of the rocket engines are small compared to the coasting times t, and if v is the coasting velocity of the rocket ship with respect to the earth, then the earthbound twin will have aged $2t$, while the traveling twin will have aged $2t\sqrt{1 - v^2/c^2}$.

This experiment is an excellent example of John Bell's statement supporting the cosmic preferred interpretation of special relativity theory. "I think the idea of an ether should be taught to students as a pedagogical device because I find there are lots of problems, which are solved more easily by imagining the existence of an ether." In order to make the analysis general for any combination of the velocity vectors of the two twins with respect to the cosmic preferred frame, the local cosmic preferred frame is oriented so that the earthbound twin is located at the origin of the xyzt frame moving with velocity V_x with respect to the XYZT frame in the positive X direction and the velocity vector of the traveling twin is located in the XY plane. The traveling twin is in the rocket ship moving with velocity $+v$ with respect to the xyzt frame on the outward journey for a time t_1 and then moves with velocity $-v$ with respect to the xyzt frame for a time $t_2 - t_1$ on the return journey.

The experiment referenced to the XYZT frame at the time when the velocity of the rocket ship is changed from $+v$ to $-v$ with respect to the xyzt frame is shown in figure 4-2.

The events referenced to the xyzt and XYZT frames are

E_0 = The start of the trip when the origins of the xyzt and XYZT frames are coincident, along with the rocket ship ($x_0 = y_0 = t_0 = 0$) ($X_0 = Y_0 = T_0 = 0$).

E_1 = The midpoint of the journey, when the velocity of the ship is changed from $+v$ to $-v$ with respect to the xyzt frame (x_1, y_1, t_1) (X_1, Y_1, T_1).

E_2 = The end of the journey when the twin in the rocket ship and the earthbound twin meet again ($x_2 = y_2 = 0$, t_2) (X_2, Y_2, T_2).

The Lorentz transformation equations are used in appendix 4C to transform the coordinates of the three events from the XYZT frame to the xyzt frame. The calculated values for the proper time measured by each of the twins verifies that the time dilation of both the rocket ship clock and the earthbound clock is caused by each of their velocities with respect to the XYZT frame. These results demonstrate that with respect to the synchronization of the cosmic preferred frame, there are no paradoxes because the proper time measured by all physical processes moving with velocity V with respect to this frame is reduced by $\sqrt{1 - V^2/c^2}$. Additionally, the proper time of each twin is less than the coordinate time of the other twin by $\sqrt{1 - V^2/c^2}$ independent of the motion of the earthbound twin with respect to the cosmic preferred frame.

4.6   The Accelerating Twin Experiment

Another example of the advantage of using the preferred frame paradigm for special relativity theory reality is demonstrated by an extension of the twin experiment. In this experiment the traveling twin is imagined to make a round trip journey to a destination two million light-years away. It is assumed that the twin is able to maintain a constant acceleration of 1 g for half of the outward journey, and then reverses the direction of the rocket thrust to slow the spaceship down to land on a planet within the Andromeda galaxy. If the return journey is made in the same fashion with 1 g accelerations, then both the traveling twin and the stay-at-home twin will experience the same one g acceleration during the total time. However, the traveling twin will have aged 56.4 years, while the earthbound twin and all things on the earth will have aged four million years.

It is to be stressed that even though the traveling twin's clock is accelerating, this is still a special relativity experiment. Marder has provided an analysis of this experiment by first describing Einstein's definition of an ideal clock and then defining the parameters of the uniform acceleration of the spaceship.[4-15]

> In effect, Einstein's assumption was that the rate of a clock depends only on its velocity and not on its acceleration (although strictly speaking he was dealing only with the case where the speed v is constant).

The most sensible definition of relativistic uniform acceleration emerges from the following dynamical considerations. Imagine the spaceship to run its motors at a fixed setting of the controls, so that propellant is ejected at a constant rate and constant velocity relative to the ship. If, during the period of acceleration the mass of propellant used is negligible, compared with the total mass of the ship, its occupants and the remaining fuel, etc., then the motion of the ship should satisfy any reasonable criterion of uniform acceleration. It would certainly be one of constant acceleration according to Newtonian mechanics.

At any instant there is a particular inertial frame (whose speed in S is the same as that of the spaceship) in which the ship is momentarily at rest although it is accelerating. We call this a *co-moving frame*. A different co-moving frame exists for each instant, and our definition of uniform acceleration is that the observed acceleration is the same in every co-moving frame. This follows because the spaceship, its controls, and the rate of ejection of propellant appear exactly the same in each of them.

A summary of Marder's analysis is given in appendix 4D. However, it is not necessary to follow the mathematics to understand that the time dilation of the proper time of the traveling twin is calculated by using the velocity of the rocket ship with respect to the cosmic preferred frame (XYZT) for each small time interval of the experiment.

These two twin experiments demonstrate that with respect to the synchronization of the cosmic preferred frame, the proper time measured by all ideal clocks and physical processes with rates independent of acceleration (e.g., biological aging at 1 g) moving with velocity V with respect to this frame is reduced by $\sqrt{1 - V^2/c^2}$ independent of acceleration.

4.7　Hawking's Chronological Protection Conjecture

The conventional interpretation of the mathematics of special relativity theory allows for the possibility of time travel to the past. This has led to a large number of books and papers describing how this might be accomplished. However, because of the unsolvable paradoxes resulting from this (e.g., killing your own grandmother before your mother was born, etc.), Hawking proposed his chronological protection conjecture that states that time travel to the past is not possible.

The galaxy clock experiment shown in figures 4-3a and 4-3b demonstrates how the conventional interpretation of the mathematics of special relativity theory allows the possibility of time travel to the past. Figure 4-3a is a world-map of the experiment referenced to the cosmic preferred frame (XYZT) at $T_0 = 0$. Here it is assumed that clocks A and C are located in distant galaxies, while clock B is located in our Milky Way galaxy. Clocks a, b, and c are located in the xyzt frame that is moving with velocity V_x with respect to the cosmic preferred frame as shown. The proper times and the location of the clocks in the xyzt frame are copied from figure 3-1a where the XYZT frame replaces the xyzt frame and the xyzt frame replaces the x'y'z't' frame and demonstrates the effects of time synchronization and the Lorentz contraction for the frame moving with respect to the XYZT frame.

Figure 4-3b is the world-map of the experiment referenced to the moving xyzt frame at $t_0 = 0$. The proper times and the location of the clocks in the XYZT frame are copied from figure 3-1c where the XYZT frame replaces the xyzt frame and the xyzt frame replaces the x'y'z't' frame and demonstrates the effects of time synchronization and the Lorentz contraction for the cosmic preferred frame referenced to the synchronization of the xyzt frame. In the conventional interpretation of special relativity theory, figures 4-3a and 4-3b are both assumed to be valid representations of this experiment. Therefore, with respect to the synchronization of the xyzt frame shown in figure 4-3b, if an observer at the origin of the xyzt frame could travel from B to A instantaneously, he or she would have traveled back in cosmic time.

The preferred frame interpretation of the mathematics of special relativity does not allow for time travel to the past because here it is assumed that the A, B, and C galaxies are all at the same age with respect to cosmic time and the clock rates in the moving xyzt frame shown in figure 4-3a are real and caused by the coupling to the external mass/energy in the universe. There is no way the moving xyzt frame can affect the age of the A and B galaxies. Therefore, while the *effective* distance from each of the A and C galaxies to the origin of the moving xyzt frame is reduced by $\sqrt{1 - V^2/c^2}$, figure 4-3b is only an imaginary representation of reality that does not change the proper time or location of the A and C galaxies with respect to the local cosmic preferred frame. Time travel to the past is impossible because the past no longer exists.

Aerts confirms that the preferred frame interpretation of special relativity theory retains the classical view of reality by restoring "past," "present (now)," and "future," which eliminates the possibility of time travel to the past.[4-16]

> *In the classical worldview reality is . . . considered to be the "classical present." The past and the future are not real.* Although we know from Einstein's analysis of the concept of simultaneity that we cannot retain the classical view on reality as being the collection of all simultaneously happening events, there has not been proposed a real relativistic equivalence for reality in a serious way . . . *if we would interpret the effect of time dilation as a physical effect on the clocks itself, the classical view can be retained. This is exactly what the Aether theory interpretations of relativity theory propose.*

Not only does the preferred frame interpretation of special relativity return the concept of time travel to the past back to science fiction where it belongs, but it also resolves Einstein's problem of the Now.[4-17]

> *Einstein said the problem of the Now worried him seriously.* He explained that the experience of the Now means something special for man, something essentially different from the past and the future, but that this important difference does not and cannot occur within physics. That this experience cannot be grasped by science seemed to him a matter of painful but inevitable resignation. *So he concluded "that there is something essential about the Now which is just outside the realm of science."*

4.8   Newton's Bucket Experiment and Mach's Principle

Newton's rotating-bucket experiment was first documented in the *Principia* published in 1687. In this experiment it is assumed there is a water-filled bucket suspended by a rope attached to the handle. The surface of the water is flat. Then, the rope is twisted by rotating the bucket, and the bucket is released to rotate freely in the opposite direction. Initially, the water is still flat with the bucket rotating around the water. Then, because of the friction between the bucket and the water, the water will begin to rotate and its surface will become concave, resulting from the centrifugal force acting on the water.

According to the conventional interpretation of relativity theory, there would be centrifugal force even if the universe was empty.[4-18]

> If we now introduce the bucket into this empty universe, it has such a tiny mass that its presence hardly affects the shape of space at all. And so the discussion we had earlier for the bucket in special relativity applies equally well to general relativity. In contradiction to what Mach would have predicted, general relativity comes to the same answer as special relativity, and proclaims that even in an otherwise empty universe, you *will* feel pressed against the inner wall of the spinning bucket; in an otherwise empty universe, you arms *will* feel pulled outward if you spin around; in an otherwise empty universe, the rope tied between two twirling rocks *will* become taut.

However, according to Mach's principle, the dynamic force experienced by the water is caused by its motion with respect to the rest of the mass/energy in the external universe.[4-19]

> Mach, it is worth noting, was cited by Einstein as an important inspiration for the theory of relativity . . . Mach's proposal is that the water is pulled up the sides of the bucket by distant massive objects, the stars and galaxies in the universe. Somehow, when the water and distant stars are in relative motion there is an extra force, in addition to gravity, acting on the water . . . Furthermore, *Mach never worked out the details on how this newly proposed dynamic force between distant stars and the water in the bucket is supposed to work. Nobody else has either,* and this remains an impediment to Mach's case against substantival space.

The cosmic preferred frame paradigm for special relativity theory reality proposes that the effective gravitational potential resulting from all of the mass/energy in the external universe is decreased as a function of velocity with respect to the cosmic preferred frame. This suggests that this difference in gravitational potential resulting from the difference in the velocity of the rotating water as a function of the distance from the center of the rotation of the bucket is the source of the centrifugal force experienced by any gravitational mass in a rotating frame.

This centrifugal force resulting from the gravitational potential generated by all of the mass/energy in the external universe is calculated in appendix 4E by establishing the time dilation effect on a clock located a distance r from the center of the frame rotating with angular velocity ω. It is not necessary to follow the mathematics to understand that the time dilation of the clock is caused by the different gravitational potential resulting from the change in velocity as a function of distance from the center of the frame, which in turn is the cause of the centrifugal force in the rotating frame.

This verification of Mach's principle demonstrates that the centrifugal force causing the surface of the water to be concave is not a "newly proposed dynamic force between distant stars and the water in the bucket" but is the gravitational force resulting from the change in the gravitational potential from the mass/energy in the external universe as a function of velocity with respect to the cosmic preferred frame. This change in gravitational potential also causes clocks in a rotating frame to run slower as a function of distance from the center of the frame.

This analysis and the analysis of the Hafele-Keating experiment show the importance of including the gravitational time dilation effect in any real clock experiments. For example, because the gravitational time dilation effect of general relativity was not known in 1905, the following statement in Einstein's paper is incorrect.[4-20]

> Thence we conclude that a balance-clock* at the equator must go more slowly by a very small amount, than a precisely similar clock situated at one of the poles under otherwise identical conditions.
>
> *Not a pendulum-clock, which is physically a system to which the Earth belongs. This case had to be excluded.

If the earth was spherical, this would be true because the distance from the center of the earth to the equator at sea level and the distance from the center of the earth to the poles at sea level would all be equal. Therefore, gravitational time dilation would be equal for all locations on the surface of the earth at sea level and special relativity time dilation would be greater at the equator. However, because the earth is an oblate spheroid with an equal total potential surface, the special relativity time dilation at the equator is exactly cancelled by the change in the gravitational time dilation resulting from the bulge at the equator.

Therefore, the sea level clock at the equator will run exactly at the same rate as a similar clock situated at one of the poles at sea level.

Even though the concepts presented in this chapter are not testable because the mathematics is the same as the conventional interpretation of relativity theory, they are extremely important because they will change the very foundations of relativity theory and will make it consistent with quantum mechanics so that it can be comprehended by the general public.

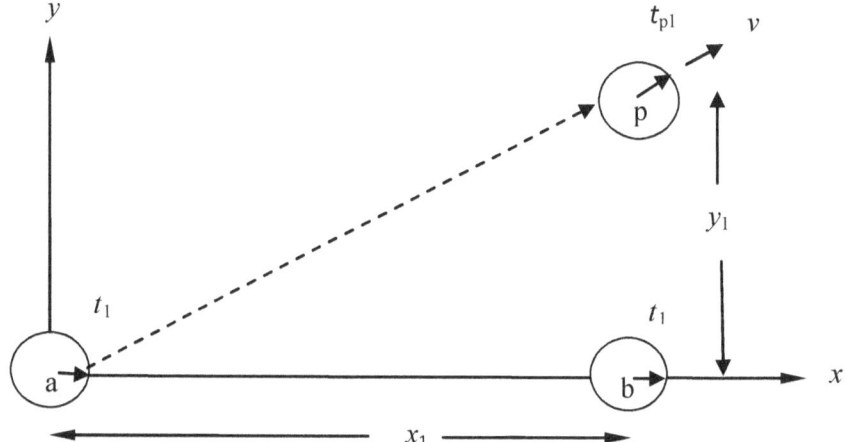

Figure 4-1a. Clock experiment 2 referenced to the xyzt frame

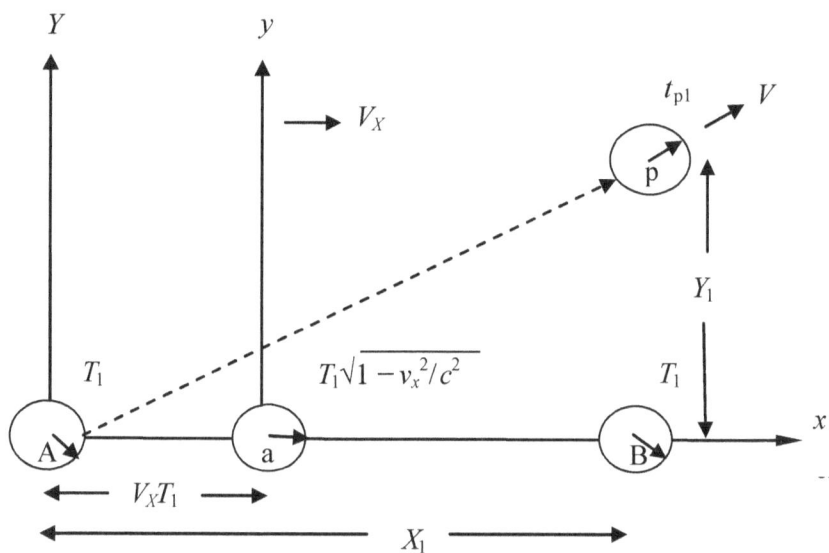

Figure 4-1b. Clock experiment 2 referenced
to the XYZT frame

Time Dilation	Δt (nsec) Eastward	Δt (nsec) Westward
Gravitational	144 ± 14	179 ± 18
Special Relativity	-184 ± 18	96 ± 10
Total	**-40 ± 23**	**275 ± 21**

Table 4-1. Predicted time differences in the
Hafele-Keating experiment[4-14]

Clock Serial No.	Δt (nsec) Eastward	Δt (nsec) Westward
120	-57	277
361	-74	284
408	-55	266
447	-51	266
Mean ± S.D.	**-59 ± 10**	**273 ± 7**

Table 4-2. Observed results of the Hafele-Keating experiment[4-14]

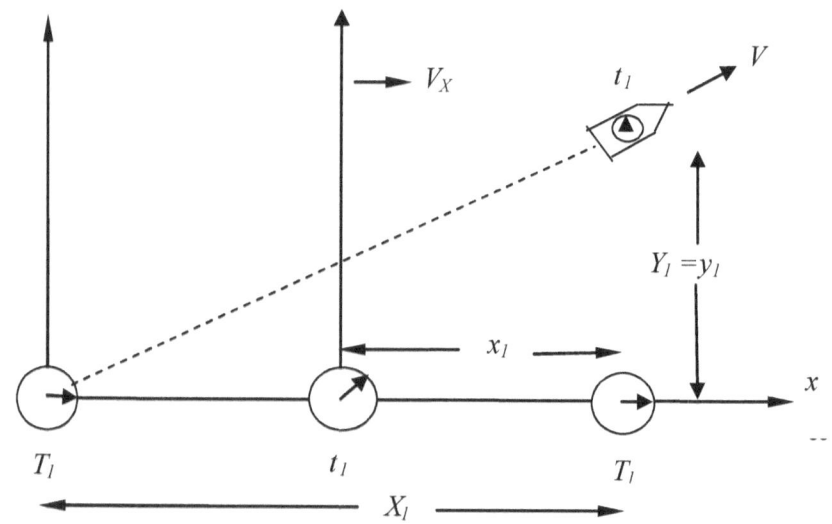

Figure 4-2. The twin experiment referenced to the
XYZT frame at turnaround (T_1).

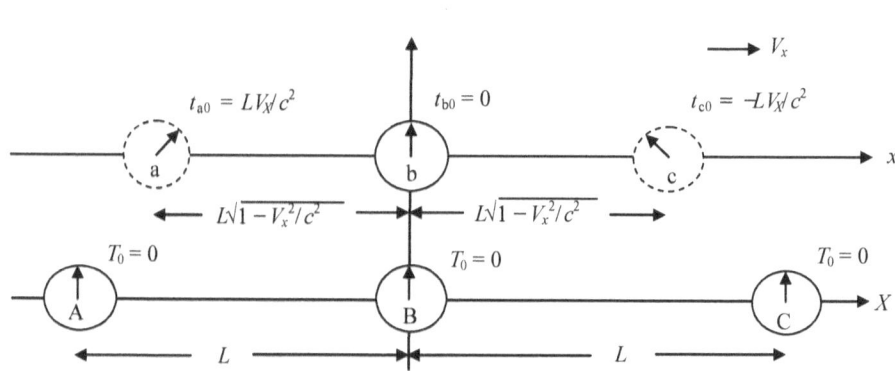

Figure 4-3a. The galaxy clock experiment
referenced to the XYZT frame at $T_0 = 0$.

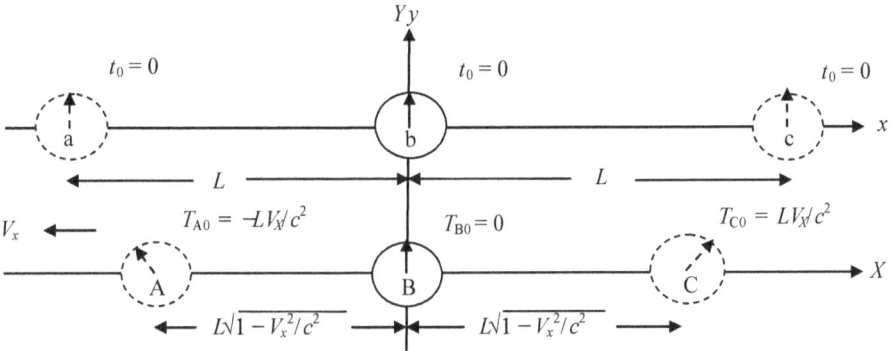

Figure 4-3b. The galaxy clock experiment
referenced to the xyzt frame at $t_0 = 0$.

5.0 John Bell's Model for Quantum Reality

An examination of the introductions of textbooks on quantum mechanics reveals the difficulty of bridging the gap from the commonsense classical physics of Newton and Maxwell to the quantum concepts developed by de Broglie, Schrödinger, Born, Bohr, Heisenberg, and Dirac. The problem has been accurately summed up by Ziock.[5-1]

> When we enter the through-the-looking-glass world of quantum mechanics, we must remember that our imagination has been molded in lifelong contact with things and events that are correctly described by Newtonian mechanics. *It will, therefore, be best if we leave behind the collection of prejudices that we sometimes fondly refer to as common sense.* Again and again, we shall have to examine with experiments the firmness of the ground on which we stand, and we shall have to entrust ourselves to the guidance of mathematics as we move about.

Examples of the difficulty of understanding the conventional interpretation of quantum mechanics that will be described in detail in the next two chapters are:

1. The double-slit experiment described in section 5.2 demonstrates the inconsistency between the wave and particle descriptions of quantum objects. In this experiment it is assumed that the act of observation can change the wave representation of a quantum object to the particle representation.
2. Schrödinger's cat experiment described in section 5-3 demonstrates the ambiguous role played by the intelligent observer in applying the mathematics of quantum mechanics.

In this experiment it is assumed that the act of observing the cat somehow modifies the actual status of the cat.

3. The dynamic parameters of quantum objects cannot be measured because the act of making a measurement with a macroscopic device changes the values of the dynamic parameters.

4. Clauser's and Aspect's experimental violation of Bell's inequality described in section 6.2 demonstrates that making a measurement of one of two phase-entangled quantum objects can immediately affect the measurement of the other object at a speed greater than the speed of light. This is inconsistent with the conventional interpretation of special relativity theory that nothing can exceed the speed of light.

Because of these problems, the developers of quantum mechanics believed that there is no quantum reality behind these mathematical rules. This has resulted in the Copenhagen interpretation of quantum mechanics.[5-2]

> The Copenhagen interpretation, physics' orthodox stance, is the way we teach and use quantum theory . . . *In the standard version of Copenhagen, observation creates the physical reality of the microscopic world, but the "observer" can, for all practical purposes, be considered to be the macroscopic measuring device, a Geiger counter, for example.*

> Copenhagen addresses the quantum enigma by telling us to pragmatically use quantum physics for the microworld and to use classical physics for the macroworld. Since we supposedly never see the microworld "directly," we can just ignore its weirdness, and thus ignore physics' encounter with consciousness. However, as quantum weirdness is seen with larger and larger objects, the ignoring gets harder, and other interpretations proliferate.

A more blunt description of the Copenhagen interpretation of quantum physics is[5-3]

> The Copenhagen interpretation was recently summarized as "*Shut up and calculate!*" That's blunt, but not completely unfair. It is, in fact, the right injunction for most physicists most of the time.

However, John Bell was one of the few physicists after Einstein who was concerned with the reality behind the mathematical models of modern physics. The quantum reality paradigm presented here is a summary of the concepts presented in his book *Speakable and unspeakable in quantum mechanics*. The basic elements of this description of Bell's quantum reality model presented in this chapter are:

1. De Broglie's original special relativity derivation of the wavelength and frequency for micro-world objects is used to describe the reality behind the de Broglie-Bohm pilot wave model of quantum reality.[5-4]

   Why is the pilot wave picture ignored in text books? Should it not be taught, not as the only way, but as an antidote to the prevailing complacency? To show that vagueness, subjectivity, and indeterminism, are not forced on us by experimental facts, but by deliberate theoretical choice?

   I will not attempt here to answer these questions. But, since the pilot wave picture still needs advertising, I will make here another modest attempt to publicize it, hoping that it may fall into the hands of a few of the many to whom even now it will be new. *I will try to present the essential idea, which is trivially simple, so compactly, so lucidly, that even some of those who know they will dislike it may go on reading, rather than set the matter aside for another day.*

2. Bell provides a description of the double-slit experiment using the de Broglie-Bohm pilot wave model of quantum reality. This analysis demonstrates that the dynamics of quantum objects are represented simultaneously by the particle *and* wave properties rather than the particle *or* wave property assumed by the Copenhagen interpretation of quantum mechanics.[5-5]

   Is it not clear from the smallness of the scintillation on the screen that we have to do with a particle? And is it not clear, from the diffraction and interference patterns, that the motion of the particle is directed by a wave? de Broglie showed in detail how the motion of a particle, passing through just one of two holes in screen, could be influenced by waves propagating through both holes. And so influenced that the particle does not go where the waves cancel out, but is attracted to where they cooperate. *This idea seems to me so natural and simple, to resolve the wave-particle dilemma in such a clear and ordinary way, that it is a great mystery to me that it was so generally ignored.* Of the founding fathers, only Einstein thought that de Broglie was on the right lines. Discouraged, de Broglie abandoned his picture for many years. He took it up again only when it was rediscovered, and more systematically presented, in 1952, by David Bohm.

3. Bell provides a description of the Schrödinger cat experiment that removes the observer ambiguity problem.[5-6]

The only "observer," which is essential in orthodox practical quantum theory, is the inanimate apparatus, which amplifies microscopic events to macroscopic consequences. Of course this apparatus, in laboratory experiments, is chosen and adjusted by the experimenters. In this sense the outcomes of experiments are indeed dependent on the mental processes of the experimenters! But once the apparatus is in place, and functioning untouched, it is a matter of complete indifference . . . according to ordinary quantum mechanics . . . whether the experimenters stay around to watch, or delegate such "observing" to computers.

Additionally, the de Broglie-Bohm pilot wave interpretation of the dynamics of quantum objects is used to construct an approximate model of the structure of the elements in the periodic table based on H. E. White's analysis of the classical Bohr-Sommerfeld elliptical orbits of the electrons in the book *Introduction to Atomic Spectra.*

5.1   The de Broglie-Bohm Pilot Wave Model

De Broglie's original model that he used to calculate the wavelength and frequency of the pilot wave required to describe the dynamics of quantum objects, strongly suggests that this periodic component of elementary particle motion is a special relativity effect, just as the magnetic field of a current carrying wire is a special relativity effect caused by the unbalance of positive and negative charges with respect to a frame moving with respect to the wire.[5-7]

> The idea, which, in my 1923-1924 works, served as the point of departure for Wave Mechanics, was the following: since for light there exists a corpuscular aspect and a wave aspect united by the relationship Energy = *h* times frequency, where *h*, Planck's constant, enters in, it is natural to suppose that, for matter as well, there exists a corpuscular *and* a wave aspect, the latter having been hitherto unrecognized. These two aspects must be united by the general formulas in which Planck's constant figures, and must contain as special cases those relationships applicable to light.
>
> *In order to elaborate this idea, it seemed to me in 1923 that it was necessary to associate a periodic element to the corpuscular concept.*

De Broglie's derivation of the wavelength, period, and velocity of the pilot wave is given in appendix 5A. David Bohm proposed a completely realistic account of quantum mechanics by assuming that the wave defined by de Broglie's special relativity model resulted in a field that caused a periodic perturbation of the classical Newtonian path. John Bell points out that he was interested in this approach, but it was never evaluated by the conventional physics community.[5-8]

> One of the things that I specifically wanted to do was to see whether there was any real objection to this idea put forward long ago by de Broglie and Bohm that you could give a completely realistic account of all quantum phenomena. de Broglie had done that in 1927, and was laughed out of court in a way that I now regard as disgraceful because his arguments were not refuted, they were simply trampled on. Bohm resurrected that theory in 1952, and was rather ignored. *I thought that the theory of Bohm and de Broglie was in all ways equivalent to quantum mechanics for experimental purposes, but nevertheless was realistic and unambiguous.*

5.2   The Double-Slit Experiment

The advantage of the de Broglie-Bohm pilot wave interpretation of the dynamics of microworld objects over the conventional Copenhagen interpretation is demonstrated by Feynman's description of the double-slit thought experiment with electrons shown in figure 5-1[5-9]. This is a thought experiment because the wavelength of the electrons is much smaller than the resolution of the holes and the electron detector. However, this experiment is used in most physics textbooks to demonstrate the conceptual problem of assuming that quantum objects are either a particle *or* a wave configuration. In the experiment the electron gun consists of a tungsten wire heated by an electric current and surrounded by a metal box with a hole in it. It is assumed that the wire is at a negative voltage with respect to the box and electrons will be ejected from the hole in the box with approximately the same energy. A thin metal plate with two holes in it is in front of the electron gun, and behind the plate is a backstop that contains a movable detector of electrons as shown in figure 5-1.

If the detector is an electron multiplier connected to a loudspeaker, then we will hear sharp "clicks" from the detector and all of the clicks are the same. There are no half-clicks. As the detector is moved around on the backstop, the rate of the clicks is faster or slower, but the loudness of each click is always the same. If we put two separate detectors at the plate, one or the other would click, but never both at once. From these results it is concluded that the detector detects all particles with approximately the same energy.

However, a plot of the rate of clicks as a function of the position of the detector with respect to the two holes (P_{12}) is shown in figure 5-1. This is the interference pattern of the Schrödinger wave equation. Similarly, if a phosphor-coated photographic plate is placed in front of the backstop, the density of the spots resulting from the irreversible events of photons hitting the plate will also show the interference pattern of the Schrödinger equation. Plots of the rate of the clicks as a function of the position of the detector for hole 1 with hole 2 covered (P_1) and for hole 2 with hole 1 covered (P_2) are shown in figure 5-1 for reference. Richard Feynman has accurately stated the difficulty of reconciling the interference pattern of quantum objects with our sense experience–based paradigm for external reality.[5-9]

> How can such an interference come about? Perhaps we should say: "Well, that means, presumably, that it is *not true* that the lumps go either through hole 1 or hole 2, because if they did, the probabilities should add. Perhaps they go in a more complicated way. They split in half and . . ." But no! They cannot, they always arrive in lumps . . . "Well, perhaps some of them go through 1, and then they go around through 2, and then around a few more times, or by some other complicated path . . . then by closing hole 2, we changed the chance that an electron that *started out* through hole 1 would finally get to the backstop . . ." But notice! There are some points at which very few electrons arrive when *both* holes are open, but which receive many electrons if we close one hole, so *closing* one hole *increased* the number from the other. Notice, however, that at the center of the pattern, P_{12} is more than twice as large as $P_1 + P_2$. It is as though closing one hole *decreased* the number of electrons, which come through the other hole. It seems hard to explain *both* effects by proposing that the electrons travel in complicated paths.

> *Many ideas have been concocted to try to explain the curve for P_{12} in terms of individual electrons going around in complicated ways through the holes. None of them has succeeded. None of them can get the right curve for P_{12} in terms of P_1 and P_2.*

However, John Bell has provided the following commonsense description of the two-hole result using the pilot wave interpretation of quantum mechanics that resolves these issues.[5-10]

> In this picture the wave always goes through both slits (as is the nature of waves) and the particle goes through only one (as is the nature of particles). But the particle is guided by the wave toward places where $|\psi|^2$ is large and away from places where $|\psi|^2$ is small. And so if the plate is in position, the particle contributes a spot to the interference pattern on the plate . . . Clearly the particle pursues a bent path in the region where the wave trains interpenetrate. *It is*

> *vital here to put away the classical prejudice that a particle moves on a straight path in "field-free" space—free, that is, from fields other than the de Broglie-Bohm!*

It is to be stressed that because the electron moves on a bent path resulting from the de Broglie-Bohm field, there is no way to establish which hole is on the path of a specific electron. Additionally, once this experiment is put in place and started, the procedure will run independently of whether the observer watches the photographic plate or not. Bell also pointed out that the number of clicks or the density of the spots on a specific location on the photographic plate is not just a measurement of the parameters of the captured electrons, but is a combination of the parameters of the electrons *and* the parameters of the macroworld objects of the experiment.[5-11]

> I suspect that they were misled by the pernicious misuse of the word "measurement" in contemporary theory. This word very strongly suggests the ascertaining of some preexisting property of some thing, any instrument involved playing a purely passive role. Quantum experiments are just not like that, as we learned especially from Bohr. *The results have to be regarded as the joint product of "system" and "apparatus," the complete experimental setup.* But the misuse of the world "measurement" makes it easy to forget this and then to expect that the "results of measurements" should obey some simple logic in which the apparatus is not mentioned. The resulting difficulties soon show that any such logic is not ordinary logic.

5.3   Schrödinger's Cat Experiment

Another example of the difficulty of reconciling the conventional interpretation of the mathematics of quantum mechanics with our sense experience–based model of external reality is demonstrated by Schrödinger's cat experiment. This experiment consists of a box containing a radioactive source, a detector that records the presence of radioactive particles, a glass bottle containing a poison, and a live cat. The detector is configured so that if one of the atoms in the radioactive source decays, then the detector will break the bottle releasing the poison that kills the cat. At the start of the experiment the box is closed, and the radioactive source is turned on for a time interval that results in a fifty-fifty chance that one of the atoms in the radioactive source decays. If the radioactive decay occurs during this period, the detector will break the bottle releasing the poison and the cat will die. However, we have no way to detect whether the cat is *alive* or *dead* until we open the box.

John Gribbon has described the Copenhagen interpretation of this experiment, where it is assumed that the cat is in a superposition state of being alive or dead until the box is opened.[5-12]

> According to . . . the strict Copenhagen interpretation, the equal probabilities for radioactive decay and no radioactive decay should produce a superposition of states. The whole experiment, cat and all, is governed by the rule that the superposition is "real" until we look at the experiment and that only at the instant of observation does the wave function collapse into one of the two states. *Until we look inside, there is a radioactive sample that has both decayed and not decayed, a glass vessel of poison that is neither broken nor unbroken, and a cat that is both dead and alive, neither alive nor dead.*

The error in this conventional interpretation of quantum mechanics results from the confusion between the *real* irreversible events in the classical macroscopic region at the interface device and the *imaginary* events in the classical macroscopic region based on the information received by the intelligent observer. In this experiment, if one of the atoms in the radioactive material decays, this results in a real irreversible event at the detector. Before that, with respect to the detector, there is only the probability of a decay of one of the atoms in the radioactive material. If a radioactive atom decays, then this results in an irreversible event when the detector records a particle and the probability function representing the decay collapses. It is important to realize that the probability function collapses with respect to the classical macroscopic region even if there is no intelligent observer or camera to record the event. Furthermore, if the time interval for the experiment is twenty-four hours and the cat is dead when the box is opened, the actual time of the real irreversible event of the cat dying can be determined by measuring the temperatures of the cat's organs.

However, the intelligent observer will not be able to determine the status of the cat until he or she looks into the box. At that event, the probability wave for the observer will collapse because of the information received. This is an excellent example of the difference between the probability function of a *real* irreversible event with respect to the classical macroscopic region at the interface device and the *imaginary* irreversible event based on the information received by the intelligent observer. John Bell has stated,[5-13]

> So I think it is not right to tell the public that a central role for conscious mind is integrated into modern atomic physics. Or that *"information" is the real stuff of physical theory*. It seems to me irresponsible to suggest that technical features of contemporary theory were anticipated by the saints of ancient religion . . . by introspection.

Another example of the difference between the real probability function of an irreversible event and the imaginary probability function of the event based on the information received by an intelligent observer that is totally in the classical macroscopic region is given by the following experiment. Imagine a box containing a roulette wheel consisting of the wheel with thirty-eight slots with the numbers 00, 0, and 1-36 and a track to move a small ball around the outside of the wheel. If the wheel and ball are set in motion and the box is closed, then the real probability function and the imaginary probability function of the intelligent observer will each be the superposition of the equal probability of the thirty-eight numbers. However, when the ball slows down and drops into one of the slots, the real probability function will collapse to the selected number even if there is no record of it, while the imaginary probability function of the intelligent observer will still be the superposition of the equal probability of the thirty-eight numbers. Then, when the intelligent observer looks inside the box, his or her imaginary probability function will collapse to the selected number.

These two examples demonstrate that the event of looking into the box does not affect the state of either the cat or the ball and wheel with respect to the physical reality inside the box.

5.4   Atomic Structure and the Periodic Table

One of the greatest achievements of twentieth-century physics was the development of the mathematics of quantum mechanics to *explain* the chemical properties of each of the elements in the periodic table. Unfortunately, while the description in the last two sections of Bell's approach to using the pilot wave model to explain the results of the two-slit experiment and Schrödinger's cat experiment was done with a minimum amount of mathematics, the detailed mathematics required to explain the electron configuration of the different elements is beyond the level of this presentation. However, this section was added to describe an approximate pilot wave model for the electrons in the different chemical elements that is constructed using Harvey White's particle property of the electron to define the possible classical Bohr-Sommerfeld elliptical orbits around the nucleus and Schrödinger's wave property of the electron to limit the elliptical orbits to integer values of the de Broglie-Bohm pilot wave wavelengths.[5-14] This approximate pilot wave model gives a description of the chemical properties of the elements that is consistent with our sense-based model of external reality.

The Bohr-Sommerfeld elliptical orbit of an electron of mass m and charge e moving with respect to the nucleus of an atom is shown in figure 5-2a. Here the nucleus is located at one focus of the ellipse. The wavelength, period, and velocity of de Broglie's special relativity model, derived in appendix 5A, were used by Schrödinger to establish a wave equation that describes the probability of the location of a quantum object in a confined volume. For example, by adding a potential energy field, Schrödinger's equation can be used to establish the probability of the position of an electron in an atom. Here, the time-independent Schrödinger equation defines a unique pilot wave for each of the electrons orbiting the nucleus of the atom. Each electron orbit is defined by three integer quantum numbers: (1) the integer *radial quantum number* (r = number of wavelengths in the radial path), (2) the integer *azimuthal quantum number* (k = number of wavelengths in the azmuthal path), and (3) the *total quantum number* (n). The transformation of these three quantum numbers (r, k, n) to the conventional three quantum numbers (n, k, m) that describe the states of the electrons in the atomic structure of the elements is described in appendix 5B.

White has pointed out,[5-14]

> Out of all the classically possible ellipses the quantum conditions allow only those for which the ratio of the major and minor axes is that of two integers, viz., the quantum numbers n and k.

In this approximate model, the total angular momentum of the electron is constant and is a function of k. Because of the historical development of atomic spectra generated by atoms with orbits of equal angular momentum, the pilot-wave for electrons with k integer values of 0, 1, 2, and 3 are named s, p, d, and f, respectively. The *magnetic quantum number* (m) defines the orientation of the elliptical orbit by the projection of the vector k with respect to the polar axis. This is shown in figure 5-2b for k values of 1, 2, and 3. Therefore, the number of m values for $k = 1, 2, 3, 4$ are $m = 3, 5, 7, 9$, respectively, as shown in figure 5-2b.

The electronic configurations of the chemical elements are shown in figure 5-3.[5-15] According to Pauli's exclusion principle, each allowed pilot wave must have a unique set of quantum states. In addition to the three quantum numbers (n, k, m) an electron can have a spin value of either "up" or "down." Therefore, as shown in figure 5-3, for each integer value of $n = 1$, there are two possible s electrons ($k = 0$); for $n = 2$, there are six possible p electrons ($k = 1$); for $n = 3$, there are ten possible d electrons ($k = 2$); and for $n = 4$, there are fourteen possible f electrons ($k = 3$).

The periodic table showing the chemical properties of the chemical elements is given in figure 5-4 for reference.[5-16] It is to be stressed that the chemical properties of each element are defined by the valence electrons in the outer shell of electrons as shown in figure 5-3. For example, the atomic number Z for each of the inert gasses that have their outer shells filled is defined in appendix 5C.[5-17] The approximate pilot-wave model described here "explains and agrees with the facts of chemistry," as well as providing an interpretation of the mathematics of quantum mechanics that is consistent with our sense experience–based paradigm for external reality (i.e., electrons are particles guided by pilot-waves that establish the allowed classical Bohr-Sommerfeld elliptical orbits and their orientation in space).

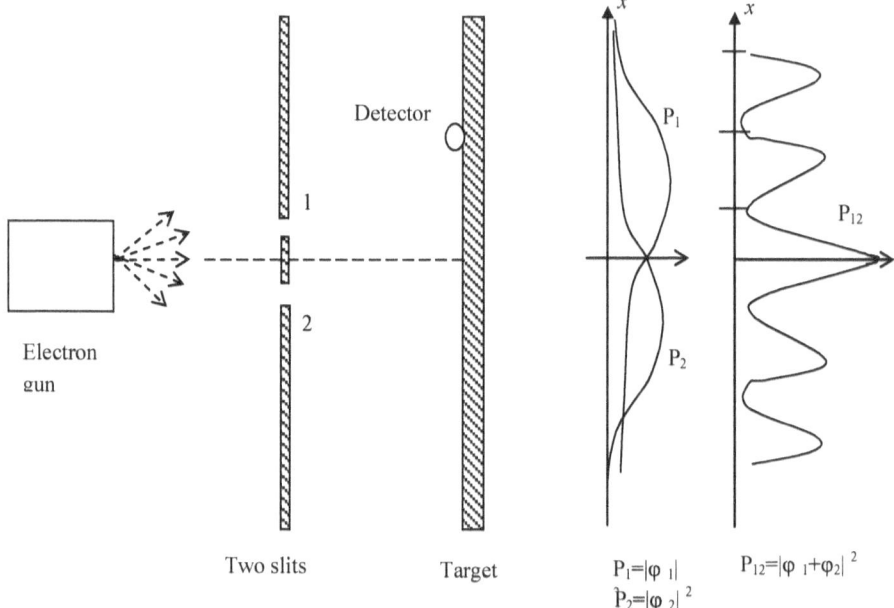

Figure 5-1. The double-slit experiment[5-9]

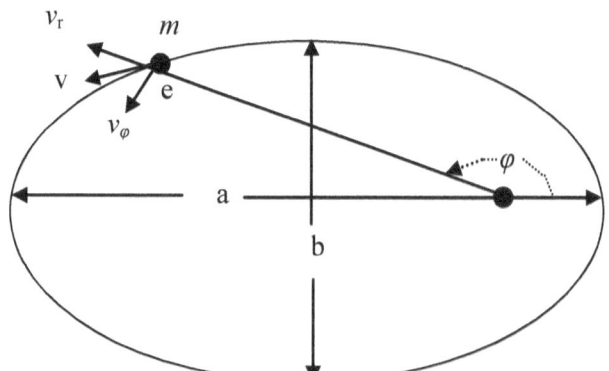

Figure 5-2a. The Bohr-Sommerfield elliptical orbit[5-14]

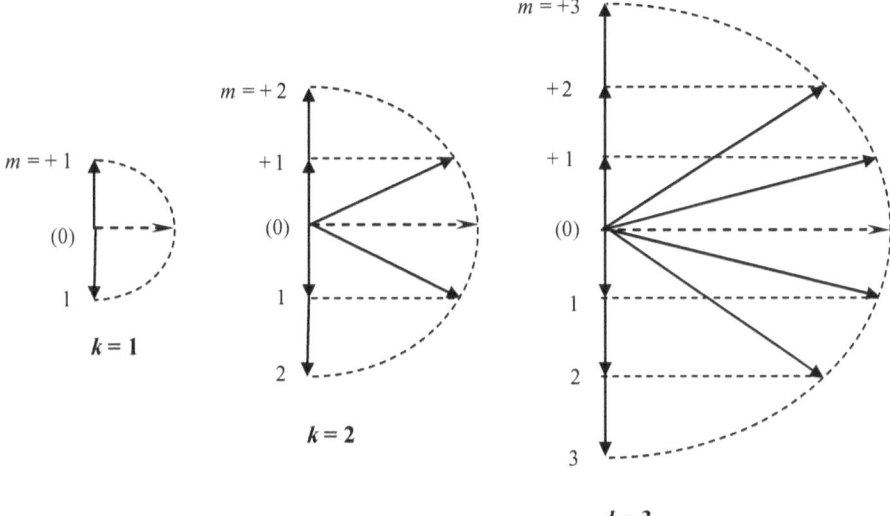

Figure 5-2b. Space quantization diagrams for
Bohr-Sommerfeld orbits $k = 1$, 2, and 3[5-14]

Figure 5-3. Electronic configuration of the elements[5-15]

Left table

Atomic No.	Element	n=1	n=2		n=3			n=4			n=5	
		s	s	p	s	p	d	s	p	d	s	p
1	H	1										
2	He	2										
3	Li	2	1									
4	Be	2	2									
5	B	2	2	1								
6	C	2	2	2								
7	N	2	2	3								
8	O	2	2	4								
9	F	2	2	5								
10	Ne	2	2	6								
11	Na	2	2	6	1							
12	Mg	2	2	6	2							
13	Al	2	2	6	2	1						
14	Si	2	2	6	2	2						
15	P	2	2	6	2	3						
16	S	2	2	6	2	4						
17	Cl	2	2	6	2	5						
18	Ar	2	2	6	2	6						
19	K	2	2	6	2	6		1				
20	Ca	2	2	6	2	6		2				
21	Sc	2	2	6	2	6	1	2				
22	Ti	2	2	6	2	6	2	2				
23	V	2	2	6	2	6	3	2				
24	Cr	2	2	6	2	6	5	1				
25	Mn	2	2	6	2	6	5	2				
26	Fe	2	2	6	2	6	6	2				
27	Co	2	2	6	2	6	7	2				
28	Ni	2	2	6	2	6	8	2				
29	Cu	2	2	6	2	6	10	1				
30	Zn	2	2	6	2	6	10	2				
31	Ga	2	2	6	2	6	10	2	1			
32	Ge	2	2	6	2	6	10	2	2			
33	As	2	2	6	2	6	10	2	3			
34	Se	2	2	6	2	6	10	2	4			
35	Br	2	2	6	2	6	10	2	5			
36	Kr	2	2	6	2	6	10	2	6			
37	Rb	2	2	6	2	6	10	2	6		1	
38	Sr	2	2	6	2	6	10	2	6		2	
39	Y	2	2	6	2	6	10	2	6	1	2	
40	Zr	2	2	6	2	6	10	2	6	2	2	
41	Cb	2	2	6	2	6	10	2	6	4	1	
42	Mo	2	2	6	2	6	10	2	6	5	1	
43	Tc	2	2	6	2	6	10	2	6	6	1	
44	Ru	2	2	6	2	6	10	2	6	7	1	
45	Rh	2	2	6	2	6	10	2	6	8	1	
46	Pd	2	2	6	2	6	10	2	6	10		
47	Ag	2	2	6	2	6	10	2	6	10	1	
48	Cd	2	2	6	2	6	10	2	6	10	2	
49	In	2	2	6	2	6	10	2	6	10	2	1
50	Sn	2	2	6	2	6	10	2	6	10	2	2
51	Sb	2	2	6	2	6	10	2	6	10	2	3
52	Te	2	2	6	2	6	10	2	6	10	2	4
53	I	2	2	6	2	6	10	2	6	10	2	5

Right table

Atomic No.	Element	n=1	n=2		n=3			n=4				n=5				n=6			n=7
		s	s	p	s	p	d	s	p	d	f	s	p	d	f	s	p	d	s
54	Xe	2	2	6	2	6	10	2	6	10		2	6						
55	Cs	2	2	6	2	6	10	2	6	10		2	6			1			
56	Ba	2	2	6	2	6	10	2	6	10		2	6			2			
57	La	2	2	6	2	6	10	2	6	10		2	6	1		2			
58	Ce	2	2	6	2	6	10	2	6	10	2	2	6			2			
59	Pr	2	2	6	2	6	10	2	6	10	3	2	6			2			
60	Nd	2	2	6	2	6	10	2	6	10	4	2	6			2			
61	Pm	2	2	6	2	6	10	2	6	10	5	2	6			2			
62	Sm	2	2	6	2	6	10	2	6	10	6	2	6			2			
63	Eu	2	2	6	2	6	10	2	6	10	7	2	6			2			
64	Gd	2	2	6	2	6	10	2	6	10	7	2	6	1		2			
65	Tb	2	2	6	2	6	10	2	6	10	9	2	6			2			
66	Dy	2	2	6	2	6	10	2	6	10	10	2	6			2			
67	Ho	2	2	6	2	6	10	2	6	10	11	2	6			2			
68	Er	2	2	6	2	6	10	2	6	10	12	2	6			2			
69	Tm	2	2	6	2	6	10	2	6	10	13	2	6			2			
70	Yb	2	2	6	2	6	10	2	6	10	14	2	6			2			
71	Lu	2	2	6	2	6	10	2	6	10	14	2	6	1		2			
72	Hf	2	2	6	2	6	10	2	6	10	14	2	6	2		2			
73	Ta	2	2	6	2	6	10	2	6	10	14	2	6	3		2			
74	W	2	2	6	2	6	10	2	6	10	14	2	6	4		2			
75	Re	2	2	6	2	6	10	2	6	10	14	2	6	5		2			
76	Os	2	2	6	2	6	10	2	6	10	14	2	6	6		2			
77	Ir	2	2	6	2	6	10	2	6	10	14	2	6	9		0			
78	Pt	2	2	6	2	6	10	2	6	10	14	2	6	9		1			
79	Au	2	2	6	2	6	10	2	6	10	14	2	6	10		1			
80	Hg	2	2	6	2	6	10	2	6	10	14	2	6	10		2			
81	Tl	2	2	6	2	6	10	2	6	10	14	2	6	10		2	1		
82	Pb	2	2	6	2	6	10	2	6	10	14	2	6	10		2	2		
83	Bi	2	2	6	2	6	10	2	6	10	14	2	6	10		2	3		
84	Po	2	2	6	2	6	10	2	6	10	14	2	6	10		2	4		
85	At	2	2	6	2	6	10	2	6	10	14	2	6	10		2	5		
86	Rn	2	2	6	2	6	10	2	6	10	14	2	6	10		2	6		
87	Fr	2	2	6	2	6	10	2	6	10	14	2	6	10		2	6		1
88	Ra	2	2	6	2	6	10	2	6	10	14	2	6	10		2	6		2
89	Ac	2	2	6	2	6	10	2	6	10	14	2	6	10		2	6	1	2
90	Th	2	2	6	2	6	10	2	6	10	14	2	6	10		2	6	2	2
91	Pa	2	2	6	2	6	10	2	6	10	14	2	6	10	2	2	6	1	2
92	U	2	2	6	2	6	10	2	6	10	14	2	6	10	3	2	6	1	2
93	Np	2	2	6	2	6	10	2	6	10	14	2	6	10	4	2	6	1	2
94	Pu	2	2	6	2	6	10	2	6	10	14	2	6	10	6	2	6		2
95	Am	2	2	6	2	6	10	2	6	10	14	2	6	10	7	2	6		2
96	Cm	2	2	6	2	6	10	2	6	10	14	2	6	10	7	2	6	1	2
97	Bk	2	2	6	2	6	10	2	6	10	14	2	6	10	8	2	6	1	2
98	Cf	2	2	6	2	6	10	2	6	10	14	2	6	10	10	2	6		2
99	Es	2	2	6	2	6	10	2	6	10	14	2	6	10	11	2	6		2
100	Fm	2	2	6	2	6	10	2	6	10	14	2	6	10	12	2	6		2
101	Md	2	2	6	2	6	10	2	6	10	14	2	6	10	13	2	6		2

Figure 5-4. The periodic table[5-16]

Period	I	II	III	IV	V	VI	VII	VIII
I	1 H 1.?							2 He 4
II	3 Li 7.6	4 Be 9	5 B 11.10	6 C 12.13	7 N 14.15	8 O 16.18.17	9 F 19	10 Ne 20.22.21
III	11 Na 23	12 Mg 24.25.26	13 Al 27	14 Si 28.29.30	15 P 31	16 S 32.34.33.36	17 Cl 35.37	18 Ar 40.36.38
IV	19 K 39.41.40	20 Ca (a)	21 Sc 45	22 Ti 47.49.50	23 V 48.46	24 Cr 52.53.50.54	25 Mn 55	26 Fe 56.54.57.58 / 27 Co 59 / 28 Ni 58.60.62.64.61
	29 Cu 63.65	30 Zn 64.66 68.67.70	31 Ga 69.71	32 Ge 74.72 70.73.76	33 As 75	34 Se (b)	35 Br 79.81	36 Kr (c)
V	37 Rb 85.87	38 Sr 86.87.84	39 Y 89	40 Zr 92.94.91.96	41 Cb 93	42 Mo (d)	43 Tc	44 Ru 102.101.104.100.99.96 / 45 Rh 103 / 46 Pd 106.108.105.110.104.102
	47 Ag 107.109	48 Cd (e)	49 In 115.113	50 Sn 120.118.116.119.117.124.122.112...	51 Sb 121.123	52 Te (g)	53 I 127	54 Xe (h)
VI	55 Cs 133	56 Ba (i)	57-71 *	72 Hf (j)	73 Ta 181	74 W 184.186.182.183.180	75 Re 187.185	76 Os 192.190 / 77 Ir 193.191 / 78 Pt 195.194.196.198.192
	79 Au 197	80 Hg (k)	81 Tl 205.203	82 Pb 206.207.204	83 Bi 209	84 Po	85 At	86 Rn 222
VII	87 Fr	88 Ra 223.224.226	89 Ac 227	90 Th 232	91 Pa 231	92 U 238.235.234	93 Np	94 Pu

(a) 40. 44. 42. 48. 43. 46
(b) 80. 78. 76. 82. 77. 74
(c) 84. 86. 82. 83. 80. 78
(d) 98. 96. 95. 92. 94. 97. 100
(e) 114. 112. 111. 110. 113. 116. 106. 108
(f) 120. 118. 116. 119. 117. 124. 122. 112. 114. 115
(g) 130. 128. 126. 125. 124. 122. 123. 120
(h) 132. 129. 131. 134. 136. 130. 128. 124. 126
(i) 138. 137. 136. 135. 134. 130. 132
(j) 180. 178. 177. 179. 176. 174
(k) 202. 200. 199. 201. 198. 204. 196

* Rare Earths

57 La	61 Il	65 Tb	69 Tm
58 Ce	62 Sm	66 Dy	70 Yb
59 Pr	63 Eu	67 Ho	71 Lu
60 Nd	64 Gd	68 Er	

Explanation.—The Roman numerals in the top row designate the "columns" of the periodic table. The roman numerals at the left give the several "periods" according to Bohr's arrangement, and the italic numerals to the upper right-hand corners of the several spaces give the ordinal numbers of the elements in these periods. Atomic numbers are indicated by bold-faced type in the upper left-hand corners. Mass numbers of the constituent isotopes are given by the numbers at the bottoms of the spaces, in the order of decreasing abundance (except for Ra).

6.0 The EPR Paradox and Bell's Inequality

In 1935 Einstein, along with Podolsky and Rosen, wrote the paper *Can Quantum-Mechanical Description of Physical Reality Be Considered Complete?* By making the two assumptions that (1) there is a reality behind the mathematical models of quantum mechanics and (2) this reality is local in that no information can travel faster than light, they concluded that the conventional interpretation of quantum mechanics is incomplete. The Einstein, Podolsky and Rosen (EPR) paper assumed that the values of momentum and position for phase-entangled quantum objects are fixed, but are hidden. This would demonstrate that the conventional interpretation of quantum mechanics is incomplete if there is a local reality for phase-entangled quantum objects (i.e., the values of these parameters must have been established at the time of generation if no "faster than light" information transfer is permitted).[6-1]

> While we have thus shown that the wave function does not provide a complete description of the physical reality, we left open the question of whether or not such a description exists. We believe, however, that such a theory is possible.

6.1 Bell's Inequality Theorem

In 1964 John Bell demonstrated that the hidden variables proposed by the EPR paper are inconsistent with the statistical predictions of quantum mechanics.[6-2]

> THE paradox of Einstein, Podolsky, and Rosen was advanced as an argument that quantum mechanics could not be a complete theory but should be supplemented by additional variables. These additional variables were to restore to the theory causality and locality. In this

note that idea will be formulated mathematically and shown to be incompatible with the statistical predictions of quantum mechanics. *It is the requirement of locality, or more precisely that the result of a measurement on one system be unaffected by operations on a distant system with which it has interacted in the past, that creates the essential difficulty.*

J. L. Townsend's thought experiment that describes Bell's inequality theorem is presented in appendix 6A,[6-3] along with an example demonstrating that quantum mechanics violates Bell's inequality in appendix 6B.[6-3] This analysis demonstrates that quantum mechanics is not consistent with a local objective reality that has no faster-than-light signaling. The experiments of Clauser and Aspect that violated Bell's inequality are described in the next section. In a BBC interview John Bell stated that a preferred frame interpretation of special relativity is the "cheapest" solution to resolve the conflict between Aspect's experimental violation of Bell's inequality and the conventional interpretation of special relativity theory.[6-4]

Narrator: "Bell's inequality is, as I understand it, rooted in two assumptions: the first is what we might call objective reality—the reality of the external world, independent of our observations; the second is locality, or non-separability, or no faster-than-light signaling. Now, Aspect's experiment appears to indicate that one of these two has to go. Which of the two would you like to hang on to?

Bell: "Well, you see, I don't really know. For me it's not something where I have a solution to sell! For me it's a dilemma. I think it's a deep dilemma, and the resolution of it will not be trivial; it will require a substantial change in the way we look at things. *But I would say that the cheapest resolution is something like going back to relativity as it was before Einstein when people like Lorentz and Poincare thought that there was an aether—a preferred frame of reference—but that our measuring instruments were distorted by motion in such a way that we could not detect motion through the aether.* Now, in that way you can imagine that there is a preferred frame of reference, and in this preferred frame of reference things do go faster than light. But then in other frames of reference when

> they seem to go not only faster than light but backwards in time, that
> is an optical illusion."

This is consistent with the preferred frame interpretation of special relativity theory described in chapter 4.

6.2.   The Experiments of Clauser and Aspect

When Bell's theorem was published in 1964, his inequality was not in a form that could be tested experimentally. However, in 1972, John Clauser, a postdoc at Berkeley, designed an experiment to determine whether the mathematics of quantum mechanics or "local reality" was correct. A simplified version of Clauser's experiment is shown in figure A-2 in the appendix and is described in the thought experiment in appendix 6B. He was motivated by his belief that local reality must be correct and quantum mechanics must be wrong. However, Rosenblum and Kuttner point out that the results of his experiment caused him to write,[6-5]

> *My own . . . vain hopes of overthrowing quantum mechanics were shattered by the data.* Confirming quantum theory's predicted violation of Bell's inequality, he showed instead that a "reasonable" description of our world, that is, a description with separability and reality, would never be possible.

The one small loophole in using Clauser's experiment to establish that the laws of physics cannot be "local" is that Clauser switched his polarizers every one hundred seconds. Therefore, it is possible that the correlated polarizers could somehow affect the measurements of the phase-entangled photons.

In 1982, Alain Aspect designed an experiment to eliminate this loophole by switching the polarizers about each 10 ns. Because this delay is small compared to L/c (40 ns) a detection event on one side and the change of orientation of the other side are separated by a spacelike interval. Aspect's experiment is described in appendix 6C. At the end of the paper describing the experiment, Aspect, Dalibard, and Roger state,[6-6]

> . . . our observed violation of Bell's inequalities indicates that the experimental accuracy was good enough for pointing out a hypothetical discrepancy with the predictions of quantum mechanics. No such effect was observed.

It is to be stressed that while the coupling between the two phase-entangled photons must transfer the measurement value of one photon at a greater-than-light-speed to the other photon, there is still no way to communicate faster than the speed of light. This is because the value of the measurement of the first photon is random and cannot be controlled.

Epilogue

The purpose of this book is to provide an interpretation of the mathematical models of relativity theory and quantum mechanics that is consistent with our sense experience–based paradigm for physical reality. The approach is to use the now-known structure of the universe to extend John Bell's preferred frame interpretation of relativity theory and the de Broglie-Bohm pilot wave interpretation of quantum mechanics. Unfortunately, acceptance of these concepts is not testable because the mathematical models are exactly the same as the conventional interpretation of modern physics.

However, the acceptance of these concepts will totally change the foundations of modern physics because it: (1) has the pedagogical advantage of restoring *cause*-and-*effect* relationships to events in the real world, (2) eliminates the possibility of time travel to the past, thus eliminating the paradoxes that have never and can never be resolved (e.g., killing one's own grandmother before your mother was born), (3) verifies Mach's principle that the centrifugal force in a rotating frame results from the motion with respect to the mass/energy in the external universe, thus eliminating the possibility of an "empty" universe, (4) demonstrates that special relativity time dilation and gravitational time dilation are the same effect when referenced to the cosmic preferred frame, (5) eliminates the conflict between the conventional interpretation of special relativity and the nonlocality of quantum mechanics described in chapter 6, and (6) restores the classical concept that the mathematics of physics describes the reality of the universe rather than merely having its predictions be in agreement with observations of the universe.

While Einstein's mathematical development of the local models of special and general relativity is correct, his assumption that there is no preferred frame in the universe is incorrect but was still a reasonable guess at the time of his death in 1955. John Bell revived the preferred frame (ether) theory because he recognized it was totally consistent with the mathematics of relativity and our classical concepts of space and time, as well as resolving the conflict between the conventional interpretation of special relativity and the nonlocality of quantum mechanics. However, Bell died in 1990 before the acceptance of the existence of the cosmic preferred frame established by the measurements of the cosmic microwave background from the big bang origin of the universe that demonstrates that he was right and Einstein was wrong.

David Albert and Rivka Galchen point out that until the past few years Bell's work and its implications have been ignored, but now these issues are "finally allowed inside the house of serious thinking about physics" [e-1]:

> *The main reaction to Bell's work—one that persists in many quarters even today—was still more obfuscation.* Bell had shown that any theory capable of reproducing the empirical predictions of quantum mechanics for entangled pairs of particles—including quantum mechanics itself—had to be genuinely physically nonlocal. *This message has been virtually ignored.*

> *The greatest worry about nonlocality, aside from its overwhelming intrinsic strangeness, has been that it intimates a profound threat to special relativity as we know it.* In the past few years this old worry—finally allowed inside the house of serious thinking about physics—has become the centerpiece of debates that may finally dismantle, distort, reimagine, solidify, or seed decay into the very foundations of physics.

The concepts presented in this book based on the observer-independent cosmic preferred frame will be instrumental in changing the very foundations of modern physics.

Appendix: Mathematical support

The purpose of this appendix is to provide the mathematical analysis that supports the cosmic preferred frame commonsense interpretation of modern physics. Most of this mathematics can be understood by anyone who has taken high school algebra.

The format used here to define the locations and the proper times of the clocks in the world-maps of experiments is to use XYZT to define the cosmic preferred frame, xyzt to define a frame moving with velocity V_X with respect to the XYZT frame, and x'y'z't' to define a frame moving with velocity v_x with respect to the xyzt frame. The events describing the dynamics of an experiment are labeled E_0, E_1 ... where the location of the event is defined by the appropriate X, Y, Z or x, y, z or x', y', z' values and the time of the event is defined by the appropriate time value (T or t or t'). Additionally, each clock is labeled with an alphabetic character that defines its rest frame (A, a, a'). Examples of the definitions of the locations and times of events are shown in table A-1. The following sections are labeled to match the chapter sections that require a mathematical description.

2A. The Lorentz contraction effect calculation

If the interferometer is moving with velocity V_X with respect to the ether frame (XYZT), then the horizontal and vertical light paths are shown by the dotted lines in figure 2-3b. If T_V is the time for light to traverse the vertical path, T_H is the time for light to traverse the horizontal path, and X_L is the length of the horizontal arm, then with respect to the XYZT frame,

$$T_V = \frac{2\sqrt{L^2 + (V_X T_V / 2)^2}}{c} = \frac{2L}{c\sqrt{1 - V_X^2/c^2}}, \qquad (2\text{-}1)$$

$$T_H = \frac{X_L}{c - V_X} + \frac{X_L}{c + V_X} = \frac{2X_L}{c(1 - V_X^2/c^2)}. \tag{2-2}$$

Because of the null result of the experiment, $T_V = T_H$. Therefore,

$$\frac{2L}{\sqrt{1 - V_X^2/c^2}} = \frac{2X_L}{c(1 - V_X^2/c^2)}. \tag{2-3}$$

Solving (2-3) for X_L,

$$X_L = \sqrt{1 - V_X^2/c^2}. \tag{2-4}$$

This is the Lorentz contraction effect of the Lorentz ether theory or special relativity theory.

2B. The time dilation effect calculation

If the interferometer is considered to be a clock, where each round trip of the vertical light beam is a "tick" of the light clock, then using (2-1) the time of each "tick" with respect to the ether frame is given by,

$$T_V = \frac{2L}{c\sqrt{1 - V_X^2/c^2}} = \frac{t_V}{\sqrt{1 - V_X^2/c^2}}. \tag{2-5}$$

Therefore,

$$t_V = T_V\sqrt{1 - V_X^2/c^2}. \tag{2-6}$$

This is the time dilation effect of the Lorentz ether theory or special relativity theory.

2C. The Lorentz transformation equations

The Lorentz transformation equations are,

$$X = \frac{x + V_X t}{\sqrt{1 - V_X^2/c^2}} \quad Y = y \quad Z = z \quad T = \frac{t + xV_X/c^2}{\sqrt{1 - V_X^2/c^2}}. \tag{2-7}$$

With reference to figure 2-3b, using the Lorentz transformation equations (2-7),

$$X_L = X_2 - V_x T_2 = \frac{x_2 + V_x t_2}{\sqrt{1 - V_x^2/c^2}} - \frac{V_x(t_2 + x_2 V_x/c^2)}{\sqrt{1 - V_x^2/c^2}}. \tag{2-8}$$

Substituting the x_2 and t_2 coordinates of event E_2 into (2-8),

$$X_L = \frac{L + V_x L/c}{\sqrt{1 - V_x^2/c^2}} - \frac{V_x(L/c + LV_x/c^2)}{\sqrt{1 - V_x^2/c^2}} = L\sqrt{1 - V_x^2/c^2}. \tag{2-9}$$

Again, this is the Lorentz contraction effect of the Lorentz ether theory or special relativity theory.

With reference to figure 2-3b, using the transformation equations and substituting the x_3 and t_3 coordinates of event E_3,

$$T_3 = \frac{t_3 + x_3 V_x/c^2}{\sqrt{1 - V_x^2/c^2}} = \frac{2L/c}{\sqrt{1 - V_x^2/c^2}} = \frac{t_3}{\sqrt{1 - V_x^2/c^2}}, \tag{2-10}$$

or,

$$t_3 = T_3 \sqrt{1 - V_x^2/c^2}. \tag{2-11}$$

Again, this is the time dilation effect.

3A. Clock experiment 1 referenced to the xyzt frame

With reference to figure 3-1a, the events describing the location in the xyzt frame ($x_{a'0}$ and $x_{c'0}$) and the time displayed by the a' and c' clocks at $t_0 = 0$ ($t'_{a'0}$ and $t'_{c'0}$) are calculated using the Lorentz transformation equations (2-7).

$$t_{a'0} = \frac{t'_{a'0} + x'_{a'0} v_x/c^2}{\sqrt{1 - v_x^2/c^2}} = 0, \tag{3-1}$$

$$t'_{a'0} = Lv_x/c^2, \tag{3-2}$$

$$x_{a'0} = \frac{x'_{a'0} + v_x\, t'_{a'0}}{\sqrt{1 - v_x^2/c^2}} - \frac{L + v_x(Lv_x/c^2)}{\sqrt{1 - v_x^2/c^2}} = -L\sqrt{1 - v_x^2/c^2}, \tag{3-3}$$

$$t_{c'0} = \frac{t'_{c'0} + x'_{c'0}\, v_x/c^2}{\sqrt{1 - v_x^2/c^2}} = 0, \tag{3-4}$$

$$t'_{c'0} = -Lv_x/c^2, \tag{3-5}$$

$$x_{c'0} = \frac{x'_{c'0} + v_x\, t'_{c'0}}{\sqrt{1 - v_x^2/c^2}} = \frac{L + v_x(-Lv_x/c^2)}{\sqrt{1 - v_x^2/c^2}} = L\sqrt{1 - v_x^2/c^2}. \tag{3-6}$$

This is shown in figure 3-1a.

With reference to figure 3-1b, the events describing the location in the xyzt frame ($x_{a'1}$ and $x_{c'1}$) and the time displayed by the a' and c' clocks at $t_1 = (L/v_x)\sqrt{1 - v_x^2/c^2}$ ($t'_{a'1}$ and $t'_{c'1}$) are calculated using the Lorentz transformation equations (2-7).

$$t_{a'1} = \frac{t'_{a'1} + x'_{a'1}\, v_x/c^2}{\sqrt{1 - v_x^2/c^2}} = (L/v_x)\sqrt{1 - v_x^2/c^2} = \frac{(L/v_x)(1 - v_x^2/c^2)}{\sqrt{1 - v_x^2/c^2}}, \tag{3-7}$$

$$t'_{a'1} = Lv_x/c^2 + L/v_x - Lv_x/c^2 = L/v_x, \tag{3-8}$$

$$x_{a'1} = \frac{x'_{a'1} + v_x\, t'_{a'1}}{\sqrt{1 - v_x^2/c^2}} = -\frac{L + v_x(L/v_x)}{\sqrt{1 - v_x^2/c^2}} = 0, \tag{3-9}$$

$$t_{b'1} = \frac{t'_{b'1} + x'_{b'1}\, v_x/c^2}{\sqrt{1 - v_x^2/c^2}} = (L/v_x)\sqrt{1 - v_x^2/c^2}, \tag{3-10}$$

$$t'_{b'1} = (L/v_x)(1 - v_x^2/c^2), \tag{3-11}$$

$$x_{b'1} = \frac{x'_{b'1} + v_x\, t'_{b'1}}{\sqrt{1 - v_x^2/c^2}} = \frac{v_x(L/v_x)(1 - v_x^2/c^2)}{\sqrt{1 - v_x^2/c^2}} = L\sqrt{1 - v_x^2/c^2}. \tag{3-12}$$

This is shown in figure 3-1b.

The coordinate time measured in the xyzt frame for clock experiment 1 is,

$$t_1 - t_0 = (L/v_x) \sqrt{1 - v_x^2/c^2}. \tag{3-13}$$

The proper time measured by clock b' for clock experiment 1 is,

$$t'_{b'1} - t'_{b'0} = (L/v_x)(1 - v_x^2/c^2) = (t_1 - t_0) \sqrt{1 - v_x^2/c^2}. \tag{3-14}$$

This is the time dilation effect of the special relativity theory, where the proper time measured by clock b' moving with velocity v_x with respect to the xyzt frame is less than the coordinate time of the xyzt frame by a factor of $\sqrt{1 - v_x^2/c^2}$.

The coordinate time measured in the x'y'z't' frame for experiment 1 is,

$$t'_{a'1} - t'_{b'0} = L/v_x. \tag{3-15}$$

The proper time measured by clock b for clock experiment 1 is,

$$t_{b1} - t_{b0} = (L/v_x) \sqrt{1 - v_x^2/c^2} = (t'_{a'1} - t'_{b'0}) \sqrt{1 - v_x^2/c^2}. \tag{3-16}$$

This is the time dilation effect of the special relativity theory, where the proper time measured by clock b moving with velocity v_x with respect to the x'y'z't' frame is less than the coordinate time of the reference frame by a factor of $\sqrt{1 - v_x^2/c^2}$.

3B.   Clock experiment 1 referenced to the x'y'z't' frame

With reference to figure 3-1c, the events describing the location in the x'y'z't' frame (x_{a0} and x_{c0}) and the time displayed by the a and c clocks at $t'_0 = 0$ (t'_{a0} and t'_{c0}) are calculated using the Lorentz transformation equations (2-7).

$$x_{a0} = \frac{x'_{a'0} + v_x t'_{a'0}}{\sqrt{1 - v_x^2/c^2}} = -L, \tag{3-17}$$

$$x'_{a'0} = -L\sqrt{1 - v_x^2/c^2}, \tag{3-18}$$

$$t_{a'0} = \frac{t'_{a'0} + x'_{a'0} v_x/c^2}{\sqrt{1 - v_x^2/c^2}} = \frac{v_x(-L\sqrt{1 - v_x^2/c^2})/c^2}{\sqrt{1 - v_x^2/c^2}} = -Lv_x/c^2, \tag{3-19}$$

$$x_{c'0} = \frac{x'_{c'0} + v_x t'_{c'0}}{\sqrt{1 - v_x^2/c^2}} = L, \tag{3-20}$$

$$x'_{c'0} = L\sqrt{1 - v_x^2/c^2}, \tag{3-21}$$

$$t_{c'0} = \frac{t'_{c'0} + x'_{c'0} v_x/c^2}{\sqrt{1 - v_x^2/c^2}} = \frac{v_x(L\sqrt{1 - v_x^2/c^2})/c^2}{\sqrt{1 - v_x^2/c^2}} = -Lv_x/c^2. \tag{3-22}$$

This is shown in figure 3-1c.

3C. Portable clock frame synchronization

With reference to figure 3-2a, the proper time of clock p when the p and b' clocks are coincident at the end of the experiment is calculated using the time dilation of the moving clock with respect to the x'y'z't' frame and the limit for the series expansion for $\sqrt{1 - v'^2_p/c^2}$.

$$t'_{p1} = t'_{b'1}\sqrt{1 - v'^2_p/c^2} = \lim_{v'_p \to 0} (L/v'_p)(1 - v'^2_p/2c^2 + \ldots) = t'_{b'1}. \tag{3-23}$$

Figure 3-2b establishes the three events defining the location and the proper time of the a', b', a, b, and p clocks with respect to the xyzt frame at the start of the experiment at $t_0 = 0$ (dotted lines) and at the end of the experiment t_1 (solid lines).

$E_{a'0}$ Start of the experiment where $x_{a0} = x'_{a'0} = x_{p0} = 0$ and $t_{a0} = t'_{a'0} = t_{p0} = 0$.

$E_{b'0}$ Start of the experiment where $x'_{b'0} = L$ and $t_0 = 0$.

$E_{b'1}$ End of the experiment at t_1.

The proper time with respect to the x'y'z't' frame ($t'_{b'0}$) and location with respect to the xyzt frame ($x_{b'0}$) of the b' clock at the start of the experiment are calculated using the Lorentz transformation equations (2-7).

$$t_0 = \frac{t'_{b'0} + x'_{b'0}\, v_x/c^2}{\sqrt{1 - v_x^2/c^2}} = 0, \tag{3-24}$$

$$t'_{b'0} = -Lv_x/c^2, \tag{3-25}$$

$$x_{b'0} = \frac{x'_{b'0} + v_x\, t'_{b'0}}{\sqrt{1 - v_x^2/c^2}} = \frac{L + v_x(-Lv_x/c^2)}{\sqrt{1 - v_x^2/c^2}} = L\sqrt{1 - v_x^2/c^2}. \tag{3-26}$$

The proper time ($t'_{b'1}$) and location ($x_{b'1}$) of the b' clock at the end of the experiment with respect to the xyzt frame are calculated using the Lorentz transformation equations (2-7),

$$t_1 = \frac{t'_{b'1} + x'_{b'1}\, v_x/c^2}{\sqrt{1 - v_x^2/c^2}} = \frac{t'_{b'1} + Lv_x/c^2}{\sqrt{1 - v_x^2/c^2}}. \tag{3-27}$$

Solving (3-27) for $t'_{b'1}$,

$$t'_{b'1} = -Lv_x/c^2 + t_1\sqrt{1 - v_x^2/c^2}. \tag{3-28}$$

Equation (3-28) demonstrates that the proper time of the b' clock at the end of the experiment ($t'_{b'1}$) is the sum of the b' clock offset at $t_0 = 0$ plus the time dilation resulting from the velocity of the b' clock with respect to the xyzt frame as shown in figure 3-2b.

The total distance traveled by the p clock at the end of the experiment (x_1) with respect to the xyzt frame is calculated using the Lorentz transformation equation (2-7).

$$x_1 = \frac{x'_{b'1} + v_x\, t'_{b'1}}{\sqrt{1 - v_x^2/c^2}} = \frac{L + v_x\left(-Lv_x/c^2 + t_1\sqrt{1 - v_x^2/c^2}\right)}{\sqrt{1 - v_x^2/c^2}}. \tag{3-29}$$

Simplifying (3-29),

$$x_1 = \frac{L(1 - v^2/c^2) + v_x t_1\sqrt{1 - v_x^2/c^2}}{\sqrt{1 - v_x^2/c^2}} = L\sqrt{1 - v_x^2/c^2} + v_x t_1. \tag{3-30}$$

This is the distance displayed in figure 3-2b.

The velocity of clock p with respect to the xyzt frame for the limit $v'_p \to 0$ is,

$$v_p = \lim_{v'_p \to 0} \frac{(v_x + v'_p)}{1 + v_x v'_p /c^2} = v_x + v'_p (1 - v_x^2/c^2). \qquad (3\text{-}31)$$

Using (3-31), the series expansion for the square root function, and the time dilation effect, the time displayed by the p clock with respect to the xyzt frame at the end of the experiment is,

$$t_{p1} = t_1\sqrt{1 - v_p^2/c^2} = t_1 \, (1 - v_x^2 / 2c^2 - v'_p v_x (1 - v_x^2/c^2) /c^2 \ldots). \qquad (3\text{-}32)$$

Using $x_1 = v_p t_1$ and (3-30) into (3-32) and simplifying,

$$t_{p1} = t_1\sqrt{1 - v_x^2/c^2} - [t_1 v_p'(1 - v_x^2/c^2)v_x/c^2] = (t_1 - Lv_x/c^2)\sqrt{1 - v_x^2/c^2}. \qquad (3\text{-}33)$$

Therefore, equations (3-28) and (3-33) can be combined for $v'_p \to 0$ and $v_x^2/c^2 \ll 1$,

$$\lim t_{p1} = t_1 \sqrt{1 - v_x^2/c^2} - (Lv_x/c^2)\sqrt{1 - v_x/c^2} = t'_{b'1}. \qquad (3\text{-}34)$$

Equation (3-34) demonstrates that a slowly moving portable clock can be used to synchronize a reference frame instead of Einstein's light beam synchronization.

4A.   Clock experiment 2

Clock experiment 2 is defined in figures 4-1a and 4-1b. Because of the time dilation effect, the proper time of the p clock with respect to the coordinate time of the xyzt frame at event E_1 is

$$t_{p1} = t_1\sqrt{1 - (x_1^2 + y_1^2)/t_1^2 c^2} \, . \qquad (4\text{-}1)$$

The Lorentz transformation equations can be used to calculate the proper time of the p clock with respect to the coordinate time of the

XYZT frame at event E_1 by first solving (2-7) for x and t as functions of X and T.

$$x = \frac{X - V_X T}{\sqrt{1 - V_x^2/c^2}} \qquad y = Y \qquad z = Z \qquad t = \frac{T - XV_X/c^2}{\sqrt{1 - V_x^2/c^2}}. \qquad (4\text{-}2)$$

Substituting the values for x and t into (4-1),

$$t_{p1} = t_1\sqrt{1 - (x_1^2 + y_1^2)/t_1^2 c^2} = T_1\sqrt{1 - (X_1^2 + Y_1^2)/T_1^2 c^2}. \qquad (4\text{-}3)$$

Equation (4-3) demonstrates that the time dilation effect of the proper time of a clock moving in an inertial frame that is moving with respect to the cosmic preferred frame is caused by the velocity of the clock referenced to the cosmic preferred frame, independent of the velocity of the inertial frame with respect to the cosmic preferred frame.

4B. Gravitational and special relativity time dilation

The gravitational time dilation effect of general relativity for two clocks located a distance r_1 and r_2 ($< r_1$) from a local gravitational body of mass M is given by,

$$t_{r1} = t_{r2}[1 + (GM/c^2)(1/r_2 - 1/r_1)], \qquad (4\text{-}4)$$

where G is Newton's gravitational constant, t_{r1} is the time measured by the clock a distance r_1 from the center of the local gravitational body and t_{r2} is the time measured by the clock r_2 from the center of the body. Therefore, the time t measured by a clock located a distance R from a local gravitational body is less than cosmic time T ($r_1 \rightarrow \infty$) by

$$t = \frac{T}{1 + (G/c^2)(M/R)}. \qquad (4\text{-}5)$$

Because the gravitational potential, a distance R from a local body of mass M, is equal to $-GM/R$, the gravitational potential resulting from the sum of all of the mass/energy in the universe with respect to the cosmic preferred frame can be represented by $-G\Sigma m/r$. Here the term $\Sigma m/r$ does not imply a linear summation of each of the gravitational bodies in the universe divided by its distance from the local clock, but instead represents the *effective* value resulting from all of the mass/energy in the universe, which includes all of the effects of general relativity (e.g., spacetime curvature, gravitational lensing, etc.). Therefore, because of the form of the gravitational time dilation equation (4-5), it is postulated that for weak local gravitational fields $[(G/c^2)(M/R) << 1]$ the gravitational time dilation effect can be represented by,

$$t = \frac{T}{(G/c^2)\Sigma m/r + (G/c^2)(M/R)} , \tag{4-6}$$

where,

$$(G/c^2)\Sigma m/r = 1. \tag{4-7}$$

According to Misner, Thorne, and Wheeler, this is in accordance with general relativity where $(G/c^2)\Sigma m/r$ is defined as the "sum for inertia" with a value equal to one.[A-1]

> The "sum for inertia" must be of the order of unity. Just such a relation of approximate identity between the mass content of the universe and its radius at the phase of maximum expansion is characteristic feature of the Friedman model and other simple models of a closed universe (chapters 27 and 30). In this respect, Einstein's theory of Mach's principle exhibits a satisfying degree of self-consistency.

> At phases of the dynamics of the universe other than the stage of maximum expansion, $r_{universe}$ can become arbitrarily small compared to $m_{universe}$. Then the ratio (21.160) can depart by powers of ten from unity. Regardless of this circumstance, *one has no option but to understand that the effective value of the "sum of inertia" is still unity after all corrections have been made for the dynamics of contraction or expansion, for retardation, etc.*

An estimate of the density of the universe based on the approximate classical model of the universe described in section 4.3 is

$$\rho = \frac{m_u}{(4/3)\pi r_u^3} = \frac{2.0 \times 10^{55}}{(4/3)\pi(1.2 \times 10^{28})^3} = 2.8 \times 10^{-30} \text{ g.cm}^{-3}, \tag{4-8}$$

where m_u is the total mass of the universe model in grams and r_u is the radius of the universe model in centimeters as shown in table A-2.

The contribution to the value of "the sum for inertia" for the earth, Sun, the Milky Way and Andromeda galaxies, a local sphere of radius one hundred million light-years, and the total universe are shown in table A-2. The total universe value of 0.20 for this approximate classical model of the total universe is in reasonable agreement with the relativistic cosmological model and is consistent with equation (4-7).

Because of the form of the special relativity time dilation equations (4-3) and (4-7), it is postulated that special relativity time dilation can be represented by

$$t = \frac{T}{(G/c^2)(\Sigma m/r) / \sqrt{1 - V^2/c^2}} . \qquad (4\text{-}9)$$

Equations (4-6) and (4-9) are a quantitative representation of the fact that the total gravitational potential experienced by a portable clock is independently decreased by being located close to a local gravitation body (gravitational time dilation) or by moving with respect to the cosmic preferred frame (special relativity time dilation). Therefore, equations (4-6) and (4-9) can be combined to give

$$t = \frac{T}{[(G/c^2)(\Sigma m/r) + (G/c^2)M/R] / \sqrt{1 - V^2/c^2}} . \qquad (4\text{-}10)$$

4C. The analysis of the twin experiment

The twin experiment referenced to the cosmic preferred frame (XYZT) at the time when the velocity of the rocket ship is changed from $+v$ to $-v$ with respect to the xyzt frame is shown in figure 4-2. The events referenced to the xyzt and XYZT frames are

E_0 = The start of the trip when the origins of the xyzt and XYZT frames are coincident, along with the rocket ship $(x_0 = y_0 = t_0 = 0)$ $(X_0 = Y_0 = T_0 = 0)$.

E_1 = The midpoint of the journey, when the velocity of the ship is changed from $+v$ to $-v$ with respect to the xyzt frame (x_1, y_1, t_1) (X_1, Y_1, T_1).

E_2 = The end of the journey when the twin in the rocket ship and the earthbound twin meet again $(x_2 = y_2 = 0, t_2)$ (X_2, Y_2, T_2).

The three points represented by these three events define the XY plane for any orientation of the rocket ship's path with respect to the path of the earthbound twin.

The conceptual advantage of referencing local experiments to the cosmic preferred frame (XYZT) is demonstrated by the following analysis of the twin experiment. Here it is assumed that the proper time measured by all clocks and physical processes moving with velocity V with respect to the cosmic preferred frame (XYZT) will be reduced by a factor of $\sqrt{1 - V^2/c^2}$. Therefore, the proper time intervals measured between events by each of the twins can be calculated by multiplying the coordinate time intervals of the XYZT frame (cosmic time) by a factor of $\sqrt{1 - V^2/c^2}$, where V is the velocity of the twin with respect to the XYZT frame. Then the Lorentz transformation equations are used to transform the coordinates of the three events from the XYZT frame to the xyzt frame.

The proper time measured by the earthbound twin at the end of the experiment is calculated using the Lorentz transformation equations (2-7) to transform the events from the XYZT frame to the xyzt frame and simplifying.

$$t_2 = T_2\sqrt{1 - V_x^2/c^2} = \sqrt{1 - (X_2^2 / T_2^2 c^2)} = t_2. \qquad (4\text{-}11)$$

The proper time measured by the traveling twin for the outward journey is calculated using the Lorentz transformation equations (2-7) to transform the events from the XYZT frame to the xyzt frame and simplifying.

$$t_1' = T_1\sqrt{1 - [(X_1^2 + Y_1^2)/T_1^2]/c^2} = t_1\sqrt{1 - V^2/c^2}. \qquad (4\text{-}12)$$

Similarly, the proper time measured by the traveling twin for the return journey is calculated using the Lorentz transformation equations (2-7) to transform the events from the XYZT frame to the xyzt frame and simplifying.

$$t_2' - t_1' = (T_2 - T_1)\sqrt{1 - \{[(X_2 - X_1)^2 + (Y_2 - Y_1)^2]/(T_2 - T_1)^2\}/c^2}$$
$$= (t_2 - t_1)\sqrt{1 - v^2/c^2}. \tag{4-13}$$

These results verify that the proper time measured by all clocks and physical processes moving with velocity V with respect to the cosmic preferred frame (XYZT) are reduced by a factor of $\sqrt{1 - V^2/c^2}$ and provide a commonsense cause-and-effect relationship between the special relativity effect in local experiments and the total mass/energy in the external universe.

4D.   The analysis of the accelerating twin experiment[4-15]

It is important to realize that the earthbound twin will experience both special relativity time dilation resulting from his or her velocity with respect to the cosmic preferred frame, as well as gravitational time dilation resulting from the gravitational potential energy of the earth itself. However, these effects are so small compared to the time dilation of the rocket ship during the four million–year journey that they can be ignored and the rocket ship considered to be accelerating with respect to the cosmic preferred frame.

The special relativity time dilation of the earthbound twin results from the velocity of the earth with respect to the cosmic preferred frame that is defined in quote[4-6] (600 km/sec). Using the series expansion for the square root function,

$$\Delta t_{sr} = -(T)(V_e^2/c^2)/2 = (4.0 \times 10^6)(6.0 \times 10^5 / 3.0 \times 10^8)^2 / 2 = -8 \text{ years}. \tag{4-14}$$

The gravitational time dilation value of the earthbound twin from the gravitational potential of the earth is defined in table A-1 [(G/c^2) $(M/R) = 0.68 \times 10^{-9}$]. With respect to (4-5),

$$\Delta t_g = -T[(G/c^2)(M/R)] = 4.0 \times 10^6 [0.68 \times 10^{-9}] = -1.0 \text{ days}. \tag{4-15}$$

Based on the assumption that the velocity of the rocket ship is zero with respect to the cosmic preferred frame (XYZT) at $T = 0$, the velocity of the accelerating twin with respect to the XYZT frame is given by

$$V = \frac{aT}{\sqrt{1 + (a^2T^2)/c^2}}. \tag{4-16}$$

This demonstrates that for small values of T, V is equal to the Newtonian velocity (aT), while for large values of T the velocity approaches the speed of light c. Therefore, the proper time of the traveling twin t_1' as a function of the proper time of the cosmic preferred frame for half of the outward journey T_1 is

$$t_1' = \int_0^{T1} \sqrt{1 - V^2/c^2} \, dt = \int_0^{T1} \sqrt{\frac{1 - (a^2T^2)}{[1 + (a^2T^2)/c^2]c^2}} \, DT. \tag{4-17}$$

Rearranging,

$$t_1' = (c/a) \int_0^{T1} \frac{dT}{\sqrt{c^2/a^2 + T^2}}. \tag{4-18}$$

Integrating (4-18),

$$t_1' = (c/a) \ln [(a/c)T_1 + \sqrt{1 + (a^2/c^2)T_1^2}]. \tag{4-19}$$

Substituting $T_1 = 1 \times 10^6$ years, $a = g = 9.81$ m/sec^2, and $c = 3.0 \times 10^8$ m/sec^2,

$$t_1' = 14.1 \text{ years}. \tag{4-20}$$

By symmetry, the earthbound twin's time for the trip is $4T_1 = 4$ million years and the traveling twin's time for the trip is $4 \, t_1' = 4 \times 14.1 = 56.4$ years.

4E. The analysis of Newton's bucket experiment

To calculate the centrifugal force in a rotating frame resulting from the gravitational potential generated by all of the mass/energy in the external universe, assume a clock located a distance r from the center of a frame rotating with angular velocity ω. Then, using (4-9) the proper time of the clock will be less than cosmic time by

$$t = \frac{T}{(G/c^2)(\Sigma m/r) / \sqrt{1 - V^2/c^2}} = \frac{T\sqrt{1 - r^2\omega^2/c^2}}{(G/c^2)(\Sigma m/r)} . \qquad (4\text{-}21)$$

Therefore, the gravitational potential at the clock is

$$\varphi = - \frac{(G\Sigma m/r)}{\sqrt{1 - r^2\omega^2/c^2}} . \qquad (4\text{-}22)$$

The gravitational force acting on the clock of mass m is

$$F = -m(\partial\varphi/\partial r) = m\frac{(G/2)(2r\omega^2/c^2)\Sigma m/r}{\sqrt{(1 - r^2\omega^2/c^2)^3}}. \qquad (4\text{-}23)$$

For $(r^2\omega^2/c^2) \ll 1$ and substituting (4-7) into (4-23),

$$F = m[(G/c^2)\, \Sigma m/r]\, r\omega^2 = mr\omega^2. \qquad (4\text{-}24)$$

This is the centrifugal force equation.

5A. De Broglie's special relativity pilot wave derivation

De Broglie assumed that the energy of a particle of mass m_0 located at the origin of a frame (x'y'z't') moving with velocity v in the positive x direction with respect to an xyzt frame can be related to a periodic function that is in phase for all locations in the x'y'z't' frame.

$$E = hv'_0 = m_0 c^2, \qquad (5\text{-}1)$$

where h is Planck's constant, v'_0 is the frequency of the periodic function referenced to the x'y'z't' frame, and $E = m_0 c^2$ is Einstein's famous equation relating rest mass and energy. Therefore, the frequency is

$$v'_0 = \frac{m_0 c^2}{h}, \tag{5-2}$$

and the period with respect to the x'y'z't' frame is given by

$$t'_{p0} = \frac{1}{v'_0} = \frac{h}{m_0 c^2}. \tag{5-3}$$

This is shown in figure A-1a. Here, each revolution of the clock represents one cycle of the periodic function. It is to be stressed that the phase of the periodic element is identical for all locations in the x'y'z't' frame.

Now, de Broglie realized that because of the Lorentz transformation equations of special relativity theory, when this situation is described by the world-map of the xyzt frame, the phase of the periodic element will change with both distance and time. Therefore, he postulated that the phase difference between the clocks in the moving x'y'z't' frame and the clocks in the xyzt reference frame is equal to the phase of the pilot wave. In order to determine how the clocks of the x'y'z't' frame shown in figure A-1a will appear in the world-map of the xyzt frame at $t = 0$, de Broglie applied the Lorentz transformation equation (4-2).

$$t'_{p0} = \frac{t - vx/c^2}{\sqrt{1 - v^2/c^2}} = \frac{h}{m_0 c^2}. \tag{5-4}$$

Solving (5-4) for x, which is the wavelength of the pilot wave with respect to the xyzt frame at $t = 0$ shown in figure A-1b,

$$\lambda = x = -\frac{c^2 \sqrt{1 - v^2/c^2}}{v} \frac{h}{m_0 c^2} = -c^2 \frac{h}{v \, (m_0 / \sqrt{1 - v_x^2/c^2})c^2} = -\frac{h}{mv}, \tag{5-5}$$

where the mass with respect to the xyzt frame is defined by

$$m = m_0 / \sqrt{1 - v^2/c^2}. \tag{5-6}$$

The minus sign in equation (5-5) demonstrates that the phase of the pilot wave is earlier for increasing values of x with respect to the synchronization of the xyzt frame. This is shown in figure A-1b.

In order to calculate the period of the pilot wave, an observer at the origin of the xyzt frame can establish the phase difference between his or her clock and the succession of clocks in the x'y'z't' frame. At the origin of the xyzt frame ($x = y = z = 0$), equation (5-4) demonstrates that the period of the periodic element with respect to the xyzt frame (t_{p0}) is less than the period with respect to the x'y'z't' frame,

$$t_{p0} = t'_{p0} \sqrt{1 - v^2/c^2} \ . \tag{5-7}$$

It is important to realize that the period is smaller with respect to the proper time of the clock at the origin of the xyzt frame than it is with respect to the coordinate time of the rest frame of the particle, even though clocks in the x'y'z't' frame appear to run slower than those located in the xyzt frame. The reason for this is shown in figure A-1b. Because of the difference in the synchronization of the two frames, the clocks moving by the origin of the xyzt frame register later times than the origin clock of the xyzt frame, even though the individual clocks of the x'y'z't' frame run at a slower rate. This is an example of the effect described in chapter 3, where the proper time measured at the origin of an inertial reference frame is less than the coordinate time of an inertial frame moving with respect to the reference frame. de Broglie commented:[A-2]

> one notes the essential difference between the apparent frequency v_a of a clock in motion—a frequency which is diminished on account of the motion—and the frequency v of the wave associated with this motion which is increased by that motion. This difference between the relativistic variations of the frequency of a clock and the frequency of a wave is fundamental. It had greatly attracted my attention, and thinking over this difference determined the whole trend of my research.

As shown in figure A-1b, for every cycle of the clock at the origin of the xyzt frame the fraction ($1/n$) of the difference in two consecutive clocks in the x'y'z't' frame will be

$$\frac{1}{n} = \frac{t'_{p0} - t_{p0}}{t_{p0}} = \frac{(t_{p0} / \sqrt{1 - v_x^2/c^2} - t_{p0})}{t_{p0}} = \frac{1}{\sqrt{1 - v^2/c^2}} - 1. \quad (5\text{-}8)$$

Applying the series expansion for $1/\sqrt{1 - v^2/c^2}$,

$$\frac{1}{n} = (1 + \frac{v^2}{2c^2} + \ldots) - 1 = \frac{v^2}{2c^2}. \quad (5\text{-}9)$$

Using (5-3), (5-6), (5-7), and (5-9), the period of de Broglie's pilot wave is then

$$n\, t_p = \frac{(2c^2)}{v^2} \frac{h}{m_0 c^2} \sqrt{1 - v^2/c^2} = \frac{2h}{mv^2}, \quad (5\text{-}10)$$

and the velocity of de Broglie's pilot wave is

$$v_{pw} = \frac{\lambda}{n\, t_p} = \frac{-h}{mv} \frac{mv^2}{2h} = -\frac{v}{2}. \quad (5\text{-}11)$$

5B.   White's derivation of k, r, and n[5-14]

Because the derivation of these three quantum numbers is beyond the scope of this presentation, an outline of White's derivation is given here. Equation (5-5) relates the momentum of the electron in an atom to the wavelength of the DeBroglie-Bohm pilot-wave.

$$\frac{mv}{h} = \frac{1}{\lambda}. \quad (5\text{-}12)$$

The relationship between the incremental azmuthal path length $(r\Delta\varphi)$ and the wavelength in the path (λ_a) is,

$$\frac{r\,\Delta\varphi}{\lambda_a} = \frac{mr^2\dot{\varphi}\,\Delta\varphi}{h} = \Delta k. \quad (5\text{-}13)$$

Therefore, the integer number of wavelengths in the azmuthal path (k) is given by

$$\oint \frac{mr^2\dot{\varphi}\,d\varphi}{h} = k. \quad (5\text{-}14)$$

Keppler's second law of planetary motion states that the angular momentum of the elliptical orbit is constant. Therefore integrating (5-14),

$$\int_0^{2\pi} mr^2\dot\varphi \, d\varphi \;=\; 2\pi \, mr^2\dot\varphi \;=\; kh. \tag{5-15}$$

The relationship between the incremental radial path length (Δr) and the wavelength in the path (λ_r) is

$$\frac{\Delta r}{\lambda_r} \;=\; \frac{m\dot r \Delta r.}{h} \tag{5-16}$$

Therefore, the integer number of wavelengths in the radial path (r) is given by,

$$\oint m\dot r \, dr \;=\; rh. \tag{5-17}$$

While the detailed mathematics required to evaluate this integral is beyond the scope of this presentation, White has provided the analysis and results of this effort.

Because k and r are integers, the total quantum number (n) is an integer defined by

$$n \;=\; k+r\,, \tag{5-18}$$

and the elliptical paths allowed by the de Broglie-Bohm pilot-wave model are represented by

$$\frac{k}{n} \;=\; \frac{b}{a}\,, \tag{5-19}$$

where a and b are the major and minor axes, respectively, of the elliptical orbits. This demonstrates that the particle property of the electron defines the classical Bohr-Sommerfeld elliptical orbits around the nucleus of the atom, while the quantum wave property of the electron limits the elliptical orbits to those with integer ratios of the major and minor axes.

5C. The atomic number Z for each of the inert gases

With reference to figure 5-3,

$$Z = 2 + (2+6) + (2+6) + (2+6+10) + (2+6+10) + (2+6+10+14), \qquad (5\text{-}20)$$

where the conventional description is given below.

$$Z = 2\,(1^2 + 2^2 + 2^2 + 3^2 + 3^2 + 4^2) = 2, 10, 18, 36, 54, 86. \qquad (5\text{-}21)$$

6A. A derivation of Bell's inequality theorem[6-3]

While Bell's original paper used electrons, J. L. Townsend's thought experiment uses a source that generates pairs of phase-entangled photons in a state with total spin zero located between two Stern-Gerlach devices (polarizers). This is shown in figure A-2. The Stern-Gerlach devices contain an internal magnet that provides the orientation of the device (magnetic vector) as well as establishes the spin of an input photon. The spin of an input photon is measured by the output being either "up" or "down." The output of each Stern-Gerlach device is sent to a photo multiplier and then to the electronic coincidence monitor where the output of each phase-entangled photon is stored as shown in figure A-2.

Because of the conservation of angular momentum, if the angle between the two Stern-Gerlach devices is zero degree, then the outputs of two photons in a spin-zero state will always be different (i.e., "up"-"down" or "down"-"up"), while if the angle between the Stearn-Gerlach devices is one hundred eighty degrees, the output will always be the same (i.e., "up"-"up" or "down"-"down"). From the rules of quantum mechanics, if the angle between the Stern-Gerlach devices is equal to θ, then the probability of the outputs being ("up"-"up") is equal to

$$P(\theta) = 1/2\,\sin^2\,(\theta/2). \qquad (6\text{-}1)$$

The rules for the EPR hidden value assumption are established by positioning the Stern-Gerlach devices to make measurements of the spin along one of three, in general, nonorthogonal, coplanar directions specified by the vectors **a, b, c**. If the spin of each of the photons is established at the time of generation, then each of the photons must belong to a definite state {+**a**, −**b**, +**c**}. In order to conserve angular momentum, if photon 1 is of the type {+**a**, −**b**, +**c**}, then photon 2 must be of the type {−**a**, +**b**, −**c**}, so that if photon 1 has its spin "up" or "down" along some axis, photon 2 will have its spin oppositely directed along the same axis. Therefore, there are eight different groups that the two photons emitted in the decay of a spin-zero particle may reside in. This is shown in table A-3.

The probability of particle 1 = +**a** and particle 2 = +**b** will be

$$P(+\mathbf{a}; +\mathbf{b}) = \frac{N_3 + N_4}{\Sigma N_i}, \qquad\qquad (6\text{-}2)$$

where N_i represents the population of the i group of photon pairs and ΣN_i represents the total population of photon pairs during the experiment. Similarly,

$$P(+\mathbf{a}; +\mathbf{c}) = \frac{N_2 + N_4}{\Sigma N_i}, \qquad\qquad (6\text{-}3)$$

$$P(+\mathbf{c}; +\mathbf{b}) = \frac{N_3 + N_7}{\Sigma N_i}. \qquad\qquad (6\text{-}4)$$

By inspection,

$$N_3 + N_4 \le (N_2 + N_4) + (N_3 + N_7). \qquad\qquad (6\text{-}5)$$

Substituting (6-2), (6-3), and (6-4) into (6-5),

$$P(+\mathbf{a}; +\mathbf{b}) \le P(+\mathbf{a}; +\mathbf{c}) + P(+\mathbf{c}; +\mathbf{b}). \qquad\qquad (6\text{-}6)$$

This is known as Bell's inequality.

It is to be stressed that even though this derivation is based on quantum effects, the inequality is only dependent on the macroscopic results of the Stern-Gerlach devices. Therefore, the analysis is based on only two assumptions: (1) the conservation of angular momentum which is shown in table A-3 and (2) the spin variables are fixed at the time the particle pair is generated. Because the conservation of angular momentum is verified by the experiment, if Bell's inequality shown in equation (6-6) is violated, then the spin variables must have been changed sometime after the particles were generated.

6B.　An example of violating Bell's inequality[6-3]

Assume the vector **c** bisects the angle between **a** and **b**. Then, using (6-1), (6-6) becomes

$$\frac{\sin^2(\theta_{ab})}{2} \leq \frac{\sin^2(\theta_{ac})}{2} + \frac{\sin^2(\theta_{cb})}{2}, \tag{6-7}$$

where θ_{ab}, θ_{ac}, θ_{cb} are the angles between vectors **a**, **b**; **a**, **c**; and **c**, **b**, respectively. For $(\theta_{ab} = 120°)$, (6-7) becomes

$$\sin^2(60°) \leq 2\sin^2(30°), \tag{6-8}$$

$$3/4 \leq 1/2. \tag{6-9}$$

Equation (6-9) demonstrates that quantum mechanics is not consistent with a local objective reality that has no faster-than-light signaling. A simplified version of Clauser's experiment is shown in figure A-2. The measured probabilities for the three angular differences violate (6-6).

6C.　Aspect's experiment[6-6]

In 1982, Alain Aspect designed an experiment to eliminate the loophole in Clauser's experiment by switching the polarizers about each 10 ns. Because this delay is small compared to L/c (40 ns), a detection event on one side and the change of orientation on the other side are separated by a spacelike interval. Aspect's experiment is shown in figure 6-2. Here each polarizer has been replaced by a switching device that selects between two polarizers in different orientations: +a and +c on side I and +b and +d on side II. If the two switches work at random and are unconnected, it is possible to write Bell's inequality in a form similar to Clauser-Horne-Shimony-Hold inequalities (CHSH).

$$S = \frac{N(a,b)}{N(\infty,\infty)} + \frac{N(c,b)}{N(\infty',\infty)} + \frac{N(c,d)}{N(\infty',\infty')} - \frac{N(a,d,)}{N(\infty,\infty')} - \frac{N(c,\infty)}{N(\infty',\infty)} - \frac{N(\infty,b).}{N(\infty,\infty)} \tag{6-10}$$

Here, $N(a,b)$, $N(c,b)$, $N(c,d)$, and $N(a,d)$ are the four coincident counting rates measured in a single run (i); $N(\infty,\infty)$, $N(\infty',\infty)$, $N(\infty',\infty')$, and $N(\infty,\infty')$ are the four coincidence rates with all of the polarizers removed in an auxiliary run (ii) (i.e., the coincidence rates for the photons counted in each of the four switching configurations); and $N(c,\infty)$ and $N(\infty,b)$ are the coincidence rates with a polarizer removed on each side in a second auxiliary run (iii).

If the hidden variables assumed in the EPR paradox are generated at the creation of the phase-entangled photons, then there are sixteen different possible states of the photon pairs. The coincident events for each term of the Clauser-Horne-Shimony-Holt inequalities equation as a function of each of the sixteen postulated EPR initial spin orientations is shown in table A-4. The first four columns define the sixteen possible EPR-postulated spin orientations for particle I. The next four columns define the coincidence events for the sixteen photon states as a function of the four polarizer directions. Here the X_i values are the fraction of each of the photon states ($\Sigma X_i = 1$). The next two columns define the coincidence events for each of the sixteen photon states as a function of the +c direction and the +b direction of the polarizers. Finally, the S column defines the contribution to the S value for the three positive terms and the three negative terms for each of the sixteen photon states. With reference to table A-4, the maximum value of S is 0 if only the eight photon types with zero in the S column are generated and the minimum value of S is equal to -1 if only the eight photon types with $-X_i$ in the S column are generated. This demonstrates that if the EPR hidden values are generated at the creation of the phase-entangled photons, the limits of S are

$$-1 \leq S \leq 0. \tag{6-11}$$

Two runs were performed in order to test the Bell inequalities. In each run the selected polarizer orientations are: $[(+a,+b) = (+c,+b) = (+c,+d) = 22.5°; (+a,+d) = 67.5°]$. The average value of S for the two runs is

$$S_{\text{expt}} = 0.101 \pm 0.020, \tag{6-12}$$

violating the inequality $S \leq 0$ by 5 standard deviations. On the other hand, for our solid angles and polarizer efficiencies, quantum mechanics predicts,

$$S_{QM} = 0.112 . \tag{6-13}$$

Event Label	Reference Frame	Clock Name	Clock Reference	Event
$x_{a'1}$	xyzt	a'	x'y'z't'	E_1
$t_{a'1}$	xyzt	a'	x'y'z't'	E_1
x'_{b2}	x'y'z't'	b	xyzt	E_2
t'_{b2}	x'y'z't'	b	xyzt	E_2
$X_{c'3}$	XYZT	c'	x'y'z't'	E_3
$T_{c'3}$	XYZT	c'	x'y'z't'	E_3
x'_{D4}	x'y'z't'	D	XYZT	E_4
x'_{D4}	x'y'z't'	D	XYZT	E_4

Table A-1. Examples of the format defining
the location and time of events

Object	Mass (kgm)	Distance (m)	$\Sigma m/r$	$(G/c^2)\, \Sigma m/r$
Earth	6.0×10^{24}	6.5×10^6	0.92×10^{18}	0.68×10^{-9}
Sun	2.0×10^{30}	1.5×10^{11}	0.13×10^{20}	0.96×10^{-8}
Milky Way	2.0×10^{41}	4.8×10^{20}	0.42×10^{21}	0.31×10^{-6}
Andromeda	2.0×10^{41}	2.1×10^{22}	0.95×10^{19}	0.70×10^{-8}
10^8 local Sphere	9.8×10^{45}	9.5×10^{23}	0.14×10^{23}	0.11×10^{-4}
Total Universe	2.0×10^{52}	1.2×10^{26}	0.26×10^{27}	0.20

Table A-2. Contributions to
"the sum of inertia" $(G/c^2)\, \Sigma m/r$

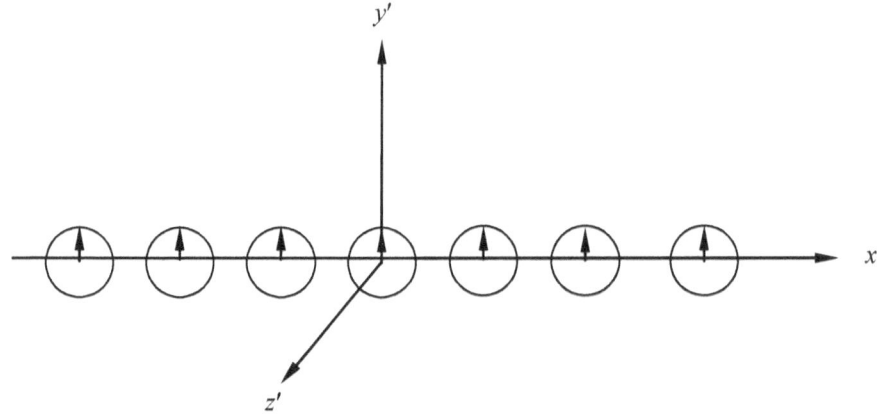

Figure A-1a. The synchronized clocks in the x′y′z′t′ frame

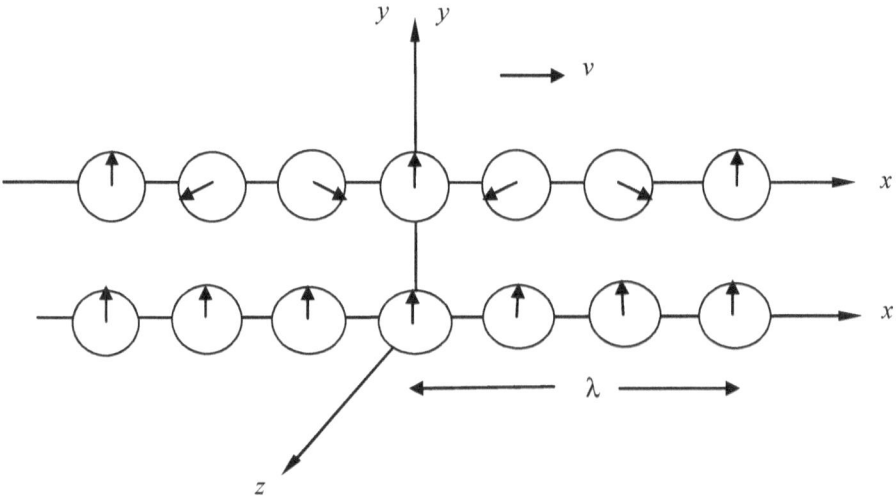

Figure A-1b. The synchronized clocks referenced
to the xyzt frame

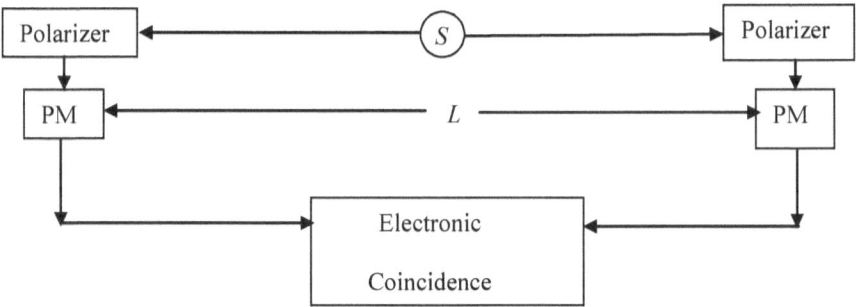

Figure A-2. The inequality experiment[6-3]

Population	Photon 1		Photon 2
N_1	**+a, +b, +c**		**−a, −b, −c**
N_2	**+a, +b, −c**		**−a, −b, +c**
N_3	**+a, −b, +c**		**−a, +b, −c**
N_4	**+a, −b, −c**		**−a, +b, +c**
N_5	**−a, +b, +c**		**+a, −b, −c**
N_6	**−a, +b, −c**		**+a, −b, +c**
N_7	**−a, −b, +c**		**+a, +b, −c**
N_8	**−a, −b, −c**		**+a, +b, +c**

Table A-3. The eight groups of spin orientations[6-3]

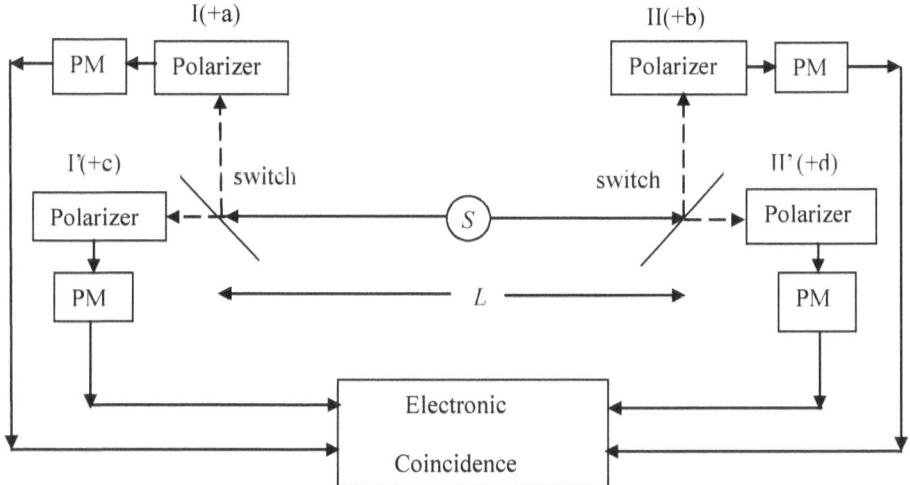

Figure A-3. Aspect's experiment[6-6]

				+Terms			− Terms			
a	b	c	d	+a;+b	+c;+b	+c;+d	+a;+d	+c	+b	S
0	0	0	0						X_0	$-X_0$
0	0	0	1						X_1	$-X_1$
0	0	1	0		X_2	X_2		X_2	X_2	0
0	0	1	1		X_3			X_3	X_3	$-X_3$
0	1	0	0							0
0	1	0	1							0
0	1	1	0			X_6		X_6		0
0	1	1	1					X_7		$-X_7$
1	0	0	0	X_8			X_8		X_8	$-X_8$
1	0	0	1	X_9					X_9	0
1	0	1	0	X_{10}	X_{10}	X_{10}	X_{10}	X_{10}	X_{10}	0
1	0	1	1	X_{11}	X_{11}			X_{11}	X_{11}	0
1	1	0	0				X_{12}			$-X_{12}$
1	1	0	1							0
1	1	1	0			X_{14}	X_{14}	X_{14}		$-X_{14}$
1	1	1	1					X_{15}		$-X_{15}$

Table A-4. Clauser-Horne-Shimony-Holt
inequalities $(-1 \leq S \geq 0)$

Endnotes

[p-1] J. S. Bell, *Speakable and unspeakable in quantum mechanics*, (Cambridge: Cambridge University Press, 1987), p. 142.

[p-2] R. Feynman, *The Character of Physical Law* (MIT Press, 1965), p. 129.

[p-3] D. Aerts, "Relativity Theory: What Is Reality?" *Foundations of Physics*, vol. 26, no. 12 (1996).

[p-4] S. Hawking and R. Penrose, *The Nature of Space and Time* (Princeton, New Jersey: Princeton University Press, 1996), pp. 3–4.

[p-5] J. S. Bell, *Speakable and unspeakable in quantum mechanics*, p. 67.

[p-6] J. S. Bell, S*peakable and unspeakable in quantum mechanics*, p. 192.

[1-1] A. Einstein, *Relativity—The Special and General Theory* (New York: Three Rivers Press, 1961), p. 81.

[1-2] J. S. Bell, S*peakable and unspeakable in quantum mechanics*, p. 170.

[2-1] H. A. Lorentz, "Electromagnetic Phenomena in a System Moving With Any Velocity Less Than That of Light," *Proceedings of the Academy of Sciences of Amsterdam* (1904). (English translation: *The Principle of Relativity*, 1923, Methuen and Co., republished by Dover, London, 1952), pp. 11–34.

[2-2] W. Rindler, *Essential Relativity, Special, General and Cosmological* (New York: Van Nostrand Reinhold, 1969), p. 48.

[3-1] A. Einstein, "On the Electrodynamics of Moving Bodies," *Annallen der Physik*, p. 891 (1905). (English translation: *The Principle of Relativity* (with notes by A. Sommerfeld), translated

by W. Perrett and G. B. Jeffery, 1923, Methuen and Co., republished by Dover, London, 1952), pp. 37–65.

[3-2]   L. Marder, *Time and the Space-Traveller* (Philadelphia: University of Pennsylvania Press, 1971), p. 24.

[3-3]   Einstein's 1905 paper.

[3-4]   M. S. Longair, *Theoretical Concepts in Physics* (Cambridge: Cambridge University Press, 1984), pp. 264–275.

[4-1]   P. C. W. Davies and J. R. Brown, *The Ghost in the Atom* (Cambridge: Cambridge University Press, 1986), p. 49.

[4-2]   L. Marder, *Time and the Space-Traveller*, p. 116.

[4-3]   L. Smolin, *The Trouble with Physics* (New York: Houghton Mifflin Company, 2006), p. 314.

[4-4]   S. Singh, *Big Bang—The Origin of the Universe* (New York, New York: Harper Collins Publishers, 2004), p. 3, p. 482.

[4-5]   J. Mather and J. Boslough, *The Very First Light* (Harper Collins Publishers, 1996), p. xviii.

[4-6]   G. Smoot and K. Davidson, *Wrinkles in Time* (Avon Books, 1993), p. 137.

[4-7]   B. Greene, *The Fabric of the Cosmos* (New York: Alfred A. Knoff, 2004), p. 228.

[4-8]   T. Moore, *A Traveler's Guide to Spacetime* (McGraw-Hill, 1995), p. 8.

[4-9]   P. Kosso, *Appearance and Reality—An Introduction to the Philosophy of Physics* (Oxford, New York: Oxford University Press, 1998), p. 78.

[4-10] J. Gott, *Time Travel in Einstein's Universe* (New York: Houghton Mifflin, 2001), pp. 69–73.

[4-11] A. Einstein, *Sidelights on Relativity* (New York: Dover Publications, 1983), p. 21.

[4-12] D. W. Sciama, *Modern Cosmology* (Great Britain: Cambridge University Press, 1971), p. 101.

[4-13] L. Smolin, *The Trouble with Physics*, p. 18.

[4-14] J. C. Hafele and R. E. Keating, "Around-the-World Atomic Clocks: Predicted   Relativistic Time Gains" and "Around-the-World Atomic Clocks: Observed Relativistic Time Gains," *Science* vol. 177 (July 1972).

[4-15] L. Marder, *Time and the Space-Traveller*, pp. 89–97.

[4-16] Aerts, "Relativity Theory: What Is Reality?" *Foundation of Physics*.

[4-17] J. Barbour, *The End of Time—The Next Revolution in Physics* (Oxford, New York: Oxford University Press, 1999), p. 143.

[4-18] B. Greene, *The Fabric of the Cosmos*, p. 74.

[4-19] P. Kosso, *Appearance and Reality—An Introduction to the Philosophy of Physics*, pp. 37–38.

[4-20] Einstein's 1905 paper.

[5-1] K. Ziock, *Basic Quantum Mechanics* (John Wiley and Sons, 1969), pp. 1–2.

[5-2] B. Rosenblum and F. Kuttner, *Quantum Enigma* (New York: Oxford Press, 2006), pp. 158–159.

[5-3] B. Rosenblum and F. Kuttner, *Quantum Enigma*, p. 112.

[5-4] J. S. Bell, *Speakable and unspeakable in quantum mechanics*, p. 160.

[5-5] J. S. Bell, *Speakable and unspeakable in quantum mechainics*, p. 191.

[5-6] J. S. Bell, *Speakable and unspeakable in quantum mechanics*, p. 170.

[5-7] L. de Broglie, *Non-Linear Wave Mechanics* (Amsterdam: Elsevier Publishing, 1960), p. 3.

[5-8] P. C. W. Davies and J. R. Brown, *The Ghost and the Atom*, p. 56.

[5-9] R. Feynman, R. Leighton, and M. Sands, *The Feynman Lectures on Physics Volume III* (Addison-Wesley Publishing, 1965), pp. 1.1–1.6.

[5-10] J. S. Bell, *Speakable and unspeakable in quantum mechanics*, pp. 112–113.

[5-11] J. S. Bell, *Speakable and unspeakable in quantum mechanics*, p. 166.

[5-12] J. Gribbon, *In Search of Schrödinger's Cat* (Bantum Books, 1984), pp. 203–205.

[5-13] J. S. Bell, *Speakable and unspeakable in quantum mechanics*, p. 170.

[5-14] H. E. White, *Introduction to Atomic Spectra* (McGraw-Hill, 1934), pp. 42–51.

[5-15] K. Ziok, *Basic Quantum Mechanics*, p. 130.

[5-16] F. K. Richtmyer and E. H. Kennard, *Introduction to Modern Physics* (McGraw-Hill, 1947), p. 738.

[5-17] F. K. Richtmyer and E. H. Kennard, *Introduction to Modern Physics*, p. 299.

[6-1] A. Einstein, B. Podolsky, and N. Rosen, "Can Quantum-Mechanical Description of Physical Reality Be Considered Complete?" *Physical Review*, vol. 47, no. 777 (1935), pp. 777–780.

[6-2] J. S. Bell, "On the Einstein, Podolsky, Rosen Paradox," *Physics*, I, pp. 195–200 (1964).

[6-3] J. L. Townsend, *A Modern Approach to Quantum Mechanics* (New York: McGraw-Hill, 1992), pp. 134–140.

[6-4] P. C. W. Davies and J. R. Brown, *The Ghost in the Atom*, pp. 48–49.

[6-5] B. Rosenblum and F. Kuttner, *Quantum Enigma*, p. 150.

[6-6] A. Aspect, J. Dalibard, and G. Roger, "Experimental Test of Bell's Inequalities Using Time-Varying Analyzers," *Physical Review Letters*, vol. 49, no. 25 (1982), pp. 1804–1807.

[e-1] D. Albert and R. Galechen, "A Quantitative Threat to Special Relativity," *Scientific American* (March 2009).

[A-1] C. W. Misner, K. S. Thorne, and J. A. Wheeler, *Gravitation* (San Francisco: W. H. Freeman, 1973), pp. 548–549.

[A-2] L. de Broglie, *Non-Linear Wave Mechanics*, p. 5.

www.ingramcontent.com/pod-product-compliance
Lightning Source LLC
Chambersburg PA
CBHW022024170526
45157CB00003B/1343